Geosynthetics from Yesterday to Today

SCIENCES

Mechanics, Field Director – Gilles Pijaudier-Cabot

Geomechanics, Subject Head –
Gioacchino Viggiani

Geosynthetics from Yesterday to Today

55 Years of French Experience

Coordinated by
Pascal Villard

iSTE

WILEY

First published 2025 in Great Britain and the United States by ISTE Ltd and John Wiley & Sons, Inc.

ISTE Ltd
27-37 St George's Road
London SW19 4EU
UK

www.iste.co.uk

John Wiley & Sons, Inc.
111 River Street
Hoboken, NJ 07030
USA

www.wiley.com

Library of Congress Control Number: 2024952939

British Library Cataloguing-in-Publication Data
A CIP record for this book is available from the British Library
ISBN 978-1-78945-175-7

ERC code:
PE1 Mathematics
PE1_7 Topology
PE10 Earth System Science
PE10_5 Geology, tectonics, volcanology

Contents

Foreword

J.P. GIROUD
Consulting Engineer and Expert, Paris, France

It gives great satisfaction to be associated with a knowledge-sharing project. The satisfaction is even greater when the project is the product of a collective endeavor. The satisfaction becomes immense when the project is the result of decades of innovation carried out within a dynamic community. And when the project is carried out within a discipline as essential as that of geosynthetics, then the sense of satisfaction doubles at the thought of the public good it serves.

This collective work has produced a book that blends theoretical and practical approaches. This book does not claim to be exhaustive but nonetheless offers a rich and amply illustrated database relevant to the activities of all types of readers. Thus, manufacturers, designers, constructors, and inspectors will find practical and well-laid out information, while teachers and students in engineering courses will find supporting material for their classes, and researchers will find inspiration for future endeavors.

This book comes at a time when geosynthetics are required to play an essential role in a century where it has become so important to protect our environment: conservation and equitable distribution of water, prevention of pollution, protection of coastlines, prevention of soil erosion, reinforcement and drainage of slopes to prevent landslides, reduction of construction material transport by promoting the use of local resources, optimal use of all kinds of materials (especially recycled products) in civil engineering projects, improvement and construction of roads, railways and waterways, and many other challenges.

This book also offers a historical perspective, as it bears witness to the extraordinary growth of geosynthetics in France for over fifty years now, in tandem with the growth of the remarkable French Committee on Geosynthetics (CFG). Indeed, this book is born out of the work of passionate engineers and innovators, some of whom contributed to the earliest rise of geosynthetics, in particular by participating in the development of these products, the refinement of techniques, the development of design methods and norms, the dissemination of knowledge, and the democratization of the various applications of geosynthetics. This long history allowed the authors to step back and take the long view. Thus, the information provided in this book is based on considerable experience. In fact, the last chapter is dedicated to experience drawn from the observation of the behavior of older structures incorporating geosynthetics constructed in France.

It is important to note that the authors of the different chapters in this book have engaged in diverse professional fields. At various points in their career they have been involved in all the activities that make up the geosynthetics discipline, namely: manufacturing geosynthetics, developing new geosynthetics, carrying out tests on geosynthetics and analyzing their results, conducting research on geosynthetics and their applications, perfecting and publishing design methods for applications of geosynthetics, integrating geosynthetics into the civil engineering syllabus, developing norms and quality-control methods related to geosynthetics, promoting the certification of geosynthetics, designing projects involving geosynthetics, installing and inspecting geosynthetics in the field, monitoring the behavior of projects that incorporate geosynthetics, evaluating the performance of geosynthetics in different kinds of projects, etc. The range of authors in this book illustrates a core feature of the discipline of geosynthetics: the intense cooperation between professionals with different training and competencies.

The authors of this book participate actively in international conferences on geosynthetics and are well aware of the treasures available in the international literature on geosynthetics. However, they found ample material in French publications to serve as sources of technical information and data for the preparation of this book initially published in French. As it provides an opportunity to a wide spectrum of readers to access the wealth of experience accumulated in France over decades of uses of geosynthetics, this English version of the book is an important contribution to the international literature on geosynthetics. This English version opens up the realm of French experience to the international community of geosynthetics. This Foreword cannot ignore this essential characteristic of the book and, naturally, mentions the French experience from the innovations of the era of pioneers to the modern achievements.

This book gives the reader a chance to review the essential role that France has played in developing the discipline of geosynthetics. The first impetus came from the industrial production in France, from 1967 onwards, of nonwoven fabrics, which would be known as 'geotextiles' ten years later. Thanks to their many applications, these materials quickly became indispensable in civil engineering projects. The role played by the manufacturers is essential: let us not forget that the geosynthetics discipline would not exist without geosynthetics and, therefore, without the manufacturers. French manufacturers must be credited, in particular, for the in-factory production of bituminous geomembranes from the 1970s onwards, which is unique on a global level, and for the development of innovative geosynthetics such as multilayer geotextile filters, drainage geocomposites, and geosynthetics equipped with sensors.

As a result of the initial impetus from the manufacturers, French engineers pioneered the design and implementation of a number of geosynthetic applications: unpaved roads stabilized using geotextiles (as soon as 1968), the first embankment on compressible soil reinforced by geotextile (1969), the first use of nonwoven geotextiles as in-soil filters (1970), the first use of a geotextile as a filter within a dam (1970), the first vertical wall reinforced with geotextiles (1971), the first use of a geotextile to protect a geomembrane (1971), the first use of a geomembrane in a double liner system (1971), the first double liner system using two geomembranes (1974), the first entirely geosynthetic double liner system using two geomembranes separated by a drainage geocomposite (1981), and the first use of a drainage geocomposite in a dam (1985). These French innovations inspired designers and builders around the world and contributed to geosynthetics revolutionizing civil engineering.

The success of the first applications of geosynthetics and the success of subsequent applications were only possible because of the quality of the geosynthetics used. This is again an area where France played an important role, being the arena where several novel laboratory tests were developed. For instance: shear tests using specially constructed shear boxes for testing interfaces containing geosynthetics; tests to measure the watertightness of geomembranes; tensile tests on wide specimens of geotextiles; and tests to assess the puncture resistance of geomembranes in contact with soil samples representative of the field conditions. Another example is the development in France of the certification processes for geosynthetics (to ensure their reliability), which are now an example for the international community. Several French experts have played, and continue to play, an important role in the development of European and international norms and testing standards governing the quality of geosynthetics.

The French have provided an intellectual impetus to the discipline of geosynthetics. Thus, French authors (both researchers and engineers) were pioneers in the development of design concepts and methods relating to several geosynthetic applications. These are, notably: stabilization of unpaved roads using geosynthetics, stability and anchorage of geosynthetics on slopes, criteria for geotextile filters, quantification of the resistance and deformation of geosynthetics that support loads over cavities and depressions, evaluation of leakage through geomembrane liners, and sizing of drainage layers associated with geomembrane liners. As early as in the 1970s, French engineers identified and defined the basic functions of geotextiles and formulated the concept of double liner systems using geomembranes, which are essential in protecting the environment. Similarly, the terminology of the geosynthetics discipline owes its character to France: the terms *géotextile* and *géomembrane*, first written in French and then adopted around the world, led naturally to the term "geosynthetics" as well as other "geo-" terms, and probably even inspired a certain "geomania" among other disciplines and other languages!

We would also be remiss in not highlighting that France had a considerable influence on the recognition of the discipline of geosynthetics and on organizing activities within this discipline. There is no dearth of examples: the first International Conference on Geotextiles (Paris, 1977), subsequently recognized as the first International Conference on Geosynthetics; the initiation of the first national committee dedicated to geotextiles (1978), the CFG, which was later expanded to include all geosynthetics; the French initiative (1982) of the creation of the International Geotextiles Society (IGS), which was officially founded in Paris (1983) and which then became the International Geosynthetics Society. At the time of writing, France is the only country, apart from the United States, to have organized two International Conferences on Geosynthetics (Paris 1977, Nice 2002).

It is clear that this book is a direct continuation of the saga of geosynthetics in France and is a testament to the vitality of the French geosynthetics community. The English version of this book is a bridge between the French experience and the international community of geosynthetics. Going beyond the immediate applications of this book, the spirit of cooperation that led to the book being created must propel future generations to continue sharing knowledge, following in the footsteps of the pioneers of the geosynthetics discipline. This will be the lasting significance of this book.

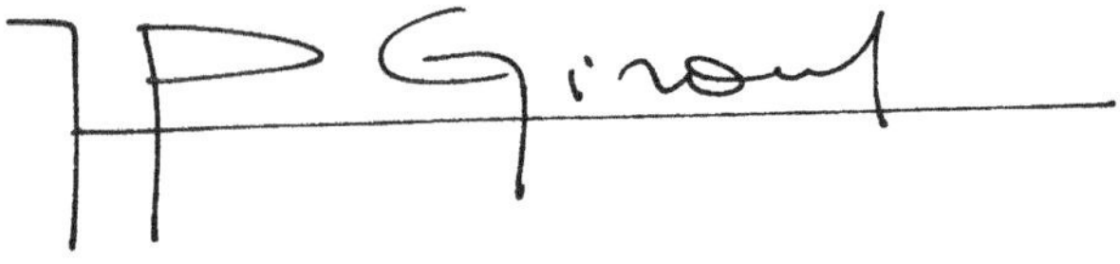

Acknowledgements

The authors would like to thank all of those who helped them with the proofreading of the English version of this book, in particular Mr Howard Murray for his expert proofreading of several chapters and Professor Marina Pantazidou for proofreading the preface.

1

Geosynthetics: Function, Properties and Characterization

Pascal VILLARD
Laboratoire 3SR, Université Grenoble Alpes, France

1.1. Definition

According to the EN ISO 10318-1 (BSI 2015a) standard, a geosynthetic (GSY) is a product, at least one of whose components is made from a synthetic or natural polymer, in the form of a sheet, a strip or a three-dimensional structure used in contact with soil and/or other materials in geotechnical and civil engineering applications.

This definition covers a very wide group of products of different appearance, whose principal functions include drainage, filtration, separation, reinforcement, protection, surface erosion-control, sealing, stabilization and crack inhibition or stress relaxation.

Sometimes used in combination with one another, these products cover a wide range of applications, considering the technical and economic advantages they afford when compared to traditional construction solutions. They are products where at least one of the components is a geosynthetic product are called "geocomposites".

The majority of geosynthetics are products manufactured from petrochemical polymers. More recently, research has been carried out around the world to respond to environmental challenges and to develop the use of recycled or even natural polymers.

Geosynthetics are divided into three categories depending on their type or their field of application:

– geotextiles (GTX) and geotextile-related products (GTP);

– geomembranes (GBR);

– geosynthetic clay barriers (GBR-C).

1.1.1. *Geotextiles and geotextile-related products*

GTX and GTP are mainly used to separate two layers of soils with different mechanical and hydraulic characteristics, filtration and drainage, reinforcement, controlling surface erosion, stabilizing and crack inhibition or stress relaxation and protecting GBR. Conventionally, GTX are permeable polymer textile products (natural or synthetic polymers: polypropylene, polyethylene, polyamide, and polyester), which can be nonwoven (GTX-NW), knitted (GTX-K) or woven (GTX-W). Woven products are made from monofilaments, multifilament yarns or tapes. Nonwoven GTX may be needle-punched or heat-bonded. Knitted GTX make it possible to join together several elements, one or more threads, filaments or other components, or even a geotextile, by interweaving them using one or more yarns. GTP are products made up of material that is planar, permeable and with a polymer base (synthetic or natural), which are used in contact with soil or other material in the field of geotechnics and civil engineering, and which do not match the definition of a geotextile. This family includes geogrids (GGR), geonets (GNT), geocells (GCE), geomats (GMA) and geospacers (GSP).

It can be seen that GTX and GTP can be made from natural materials (made up of natural fibers: jute, linen, hemp, coir, cotton, straw, etc.). They are chiefly used for separation, filtration, drainage, slope protection and erosion control or as reinforcement in temporary structures, given that they are biodegradable.

1.1.2. *Geomembranes*

Geosynthetics that offer a barrier to prevent or limit the migration of fluids (GBR, commonly called geomembranes in France) take the form of flexible sheets that are thin and impermeable, offering sealing against liquids and gases. The

EN ISO 10318-1 (BSI 2015a) standard defines terms related to geosynthetic sealing systems (DEG) as well as the structure, manufacture and assembly of GBR.

It defines:

– polymeric geosynthetic barriers (GBR-P);

– bituminous geosynthetic barriers (GBR-B);

– clay geosynthetic barriers (GBR-C) also geosynthetic clay liner (GCL).

GBR-P are made up of different polymers (high-density polyethylene (HDPE), flexible polypropylene (PPF), plasticized polyvinyl chloride (PVC-P) and ethylene propylene diene terpolymer (EPDM) and are produced by extrusion, calendering or impregnation.

GBR-B are made from a structure that is assembled in a factory. This structure is made up of geosynthetics and has the form of a sheet. The bitumen chiefly provides the sealing or barrier function.

1.1.3. *Geosynthetic clay barriers*

GBR-C, EN ISO 10318-1 (BSI 2015a), are made from a factory-assembled structure consisting of geosynthetics and in the form of a sheet in which the barrier function is essentially fulfilled by clay. It is important to note that bentonite geosynthetics only realize their sealing function after being wetted in place after they are laid, and this must be done under a confining soil layer.

1.1.4. *Standard practices*

In France, the implementation of geosynthetic solutions is supported by standard practices that guide project managers and contractors in the safe planning, designing and implementation of these different products. Geosynthetic products and their applications are governed by a quality certification that is based on standardized tests developed in partnership with the profession. The *Comité Français des Géosynthétiques* (French Geosynthetics Committee), which brings together producers, research bodies, project managers, contractors, consulting engineers, public works companies, and building contractors, is the main actor in France. This body compiles knowledge in the field, disseminates information to users, promotes the use and development of geosynthetics and new techniques, and, finally, supports standardization in France, Europe and around the world.

1.2. The various functions and applications of geosynthetics

The main functions performed by geosynthetics are as follows: separation, protection, filtration, drainage, reinforcement, sealing, erosion protection and stress relief (crack inhibition). Generally speaking, each geosynthetic performs a specific function. However, certain geosynthetics may carry out several functions simultaneously. The harmonized European standards have listed and associated the various functions of geosynthetics with the many geotechnical applications for which they are used (Table 1.1).

Function / Application	Separation	Protection	Filtration	Drainage	Reinforcement	Sealing	Anti-erosion	Anti-fissuring
Laying roads and railway lines	X		X	(1)	X			
Drainage work	X		X	X				
Riverbanks and coastlines	X		X	(1)	X		X	
Canals	X	X	X	(1)	X	X	X	
Dams and reservoirs	X	X	X	(1)	X	X	X	
Supports and foundations	X		X	(1)	X			
Tunnels		X	(1)	(1)		X		
Landfill (solid waste)	X	X	X	(1)	X	X	X	
Landfill (liquid waste)	X	X	X	(1)		X		
Pavements					X	X		X

(1) For drainage systems, refer to the application "Drainage Works".

Table 1.1. *Functions of geotextiles and geotextile-related products*

COMMENT ON TABLE 1.1.– *Functions associated with the geotechnology applications in which they may be used as per the harmonized European standards: EN 13249 (BSI 2016a), EN 13250 (BSI 2016b), EN 13251 (BSI 2016c), EN 13252 (BSI 2016d), EN 13253 (BSI 2016e), EN 13254 (BSI 2016f), EN 13255 (BSI 2016g), EN 13256 (BSI 2016h), EN 13257 (BSI 2016i), EN 13265 (BSI 2016j), EN 13361 (BSI 2018a), EN 13362 (BSI 2018b), EN 13491 (BSI 2018c), EN 13492 (BSI 2018d), EN 13493 (BSI 2018e), EN 15381 (BSI 2018f) and EN 15382 (BSI 2018g).*

1.2.1. *Separation*

GTX were initially developed to separate two soil layers with different characteristics: a subgrade from a sub-base layer or several sub-base layers with different gradings.

They allow the different layers to retain their respective properties while preventing migration or the contamination of one layer by another. Their function was therefore to prevent the mixing of adjacent layers without obstructing the movement of fluids (Figure 1.1).

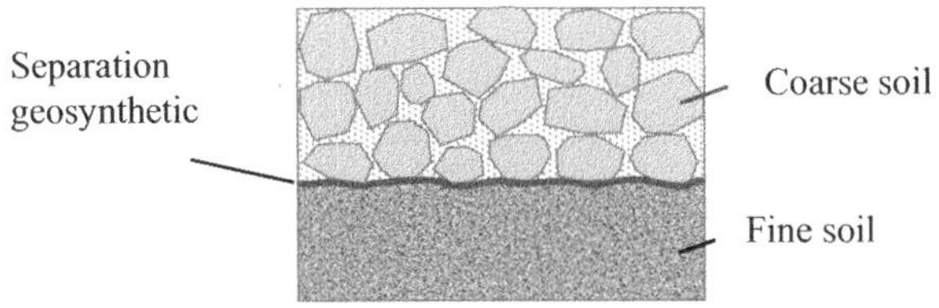

Figure 1.1. *Function – separating two layers of soil of different properties. For a color version of this figure, see www.iste.co.uk/villard/geosynthetics.zip*

These products are not subjected to significant tensile stress, but they must be capable of resisting local puncturing forces. They are designed considering their resistance to tensile stress and to puncturing, their extension under maximum load and their opening size (the dimension of the largest particle that can pass through the product).

This function is used for roadways like temporary roads or roads with low traffic values in order to minimize the formation of ruts (access roads to a work-site or forest roads), or to ensure the longevity and durability of transport infrastructure such as the base layers of roads, freeways and railways lines. They are also used in geotechnical projects when materials with very different properties have to be used together, for example, in the construction of landfills on compressible soils.

1.2.2. *Filtration*

Filtration is the geosynthetic's ability to retain soil particles whose diameter is larger than a given nominal value while allowing the passage of fluids in a direction perpendicular to its plane (Figure 1.2). This function is typically required in drainage systems, where the geosynthetic must allow water from the zone being drained to pass through and flow toward the drain, while at the same time preventing the clogging of the drain by fine particles, and preventing the progressive erosion of soil upstream of the drain.

Geotextile filters can be used in many projects, for instance, hydraulic structures (dams, basins, canals, dikes, etc.), earthwork projects, embankments on compressible soils, roadways, engineering structures, foundations and waste-storage facilities.

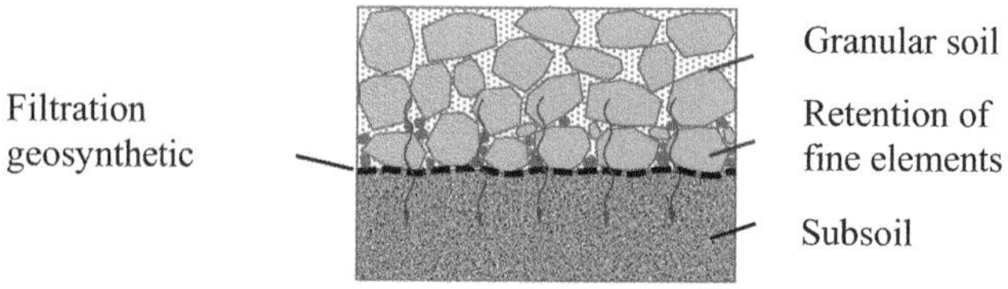

Figure 1.2. *Filtration function – retention of fine particles. For a color version of this figure, see www.iste.co.uk/villard/geosynthetics.zip*

The challenge in designing geosynthetic filters is to optimize their filtration opening sizes (in mm), which governs their ability to obstruct or allow the passage of fine particles of a given size. This will allow the passage of fluids while ensuring a specified flow rate and also preventing the migration of fine elements from one medium to another (internal erosion of the soil). It must also prevent the clogging of the product, which could in the long term make it ineffective and could compromise the durability of the structure (interstitial excess pressures could damage its stability).

1.2.3. *Drainage*

Drainage is the function of collecting fluids and transporting them toward an outlet (Figure 1.3). The transport is parallel to the plane of the geosynthetic and the fluid is captured over a large surface, which can lead to high flow rates. To avoid the drainage system being clogged and to ensure a constant flow rate, the drainage system is always used in conjunction with a filter on the side of the soil to be drained, and this retains soil particles. The drain's capacity to evacuate water in its plane is characterized by transmissivity. Drains are used in many civil engineering projects for accelerating the consolidation of soils, reducing interstitial excess pressures within structures, evacuating runoff water and leachates, and contributing to the lowering of the water table.

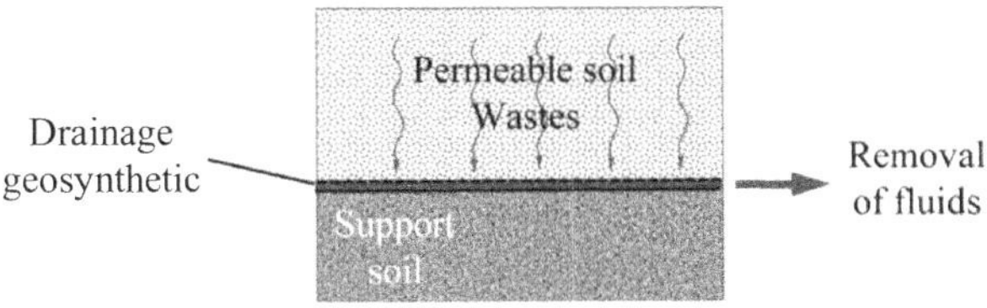

Figure 1.3. *Drainage function – collecting fluids. For a color version of this figure, see www.iste.co.uk/villard/geosynthetics.zip*

1.2.4. *Reinforcement*

Reinforcing soil is an ancient technique that consists of combining soils of mediocre quality with reinforcements whose higher tensile strength will compensate for the shortcomings of the local soil. The technique of using geosynthetics to reinforce soil has been used for many decades now in geotechnical projects. It makes it possible to both improve the stability of the structures and to permit greater deformations before failure occurs.

Because of their mechanical properties, reinforcement geosynthetics contribute to the reinforcement of soils of low bearing capacity. They act as a framework in retaining structures, add stability to sealing layers on slopes, limit the magnitude of collapses and counteract landslides. The materials chiefly used as reinforcement are woven GTX or knitted GTX that include reinforcing elements in one or more directions or open-mesh geogrids that allow greater interlocking with granular soils. Reinforcement geosynthetics are designed to resist tensile forces.

There are several ways of creating tension in the geosynthetic. In some cases, the displacement of the soil leads to tensile forces in the reinforcing material (Figure 1.4). In this case, the geosynthetic confines the soil, and this increases its strength.

In some applications, such as reinforcement over a cavity, the soil movement on the reinforcing layer during the opening of the cavity creates forces that act mainly perpendicular to the plane of the geosynthetics that curves (membrane effect), thus allowing the development of tensile forces that are opposed to the action of the collapsed soil (Figure 1.5).

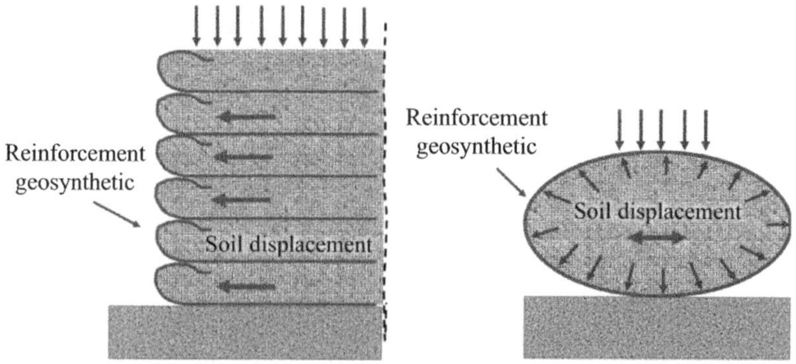

Figure 1.4. *Reinforcement through soil displacement. For a color version of this figure, see www.iste.co.uk/villard/geosynthetics.zip*

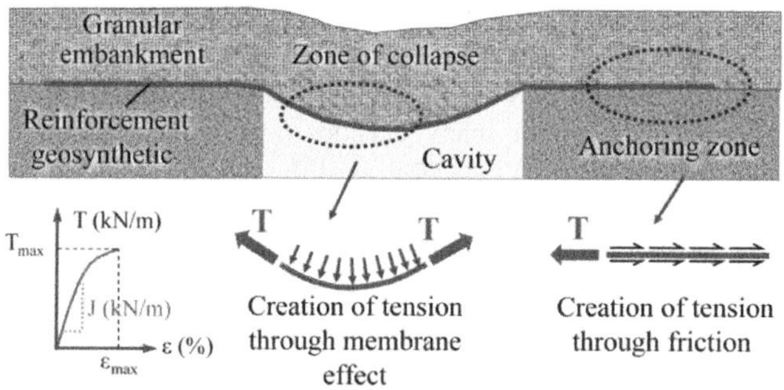

Figure 1.5. *Reinforcement geosynthetic: creating tension through a membrane effect and through friction. For a color version of this figure, see www.iste.co.uk/villard/geosynthetics.zip*

In other cases, it is the geosynthetic that is displaced with respect to the soil (e.g. anchor zone) (Figure 1.5). The relative displacement between the soil and the geosynthetic induces shear forces at the interface that create tension in the reinforcement. The optimization of the interface characteristics, which depend on the texture and roughness of the reinforcement, is essential for optimal reinforcement (a smooth interface characterized by low interface friction does not result in a significant transfer of tangential forces from the soil to the geosynthetic).

Therefore, the basic characteristics required for designing reinforcement geosynthetics are their tensile stiffness, the resistance to tension, deformation under maximum load and their interface properties (which must be determined based on the soil type or the reinforcement geosynthetic being used).

1.2.5. *Sealing*

The sealing function is essential to limit the transfer of fluids from one medium to another, especially to protect the environment from pollutants (waste storage centers) or to conserve water resources (retention basins, reservoirs and canals) (Figure 1.6). The sealing function is performed by GBR and bentonite geosynthetics. Their ability to allow a fluid to pass through in a direction perpendicular to their plane is described as "permittivity". When used in a sealing product, sealing geosynthetics are not intended to play a mechanical role and for this reason they are often used in conjunction with other geosynthetics such as protection or reinforcement geosynthetics.

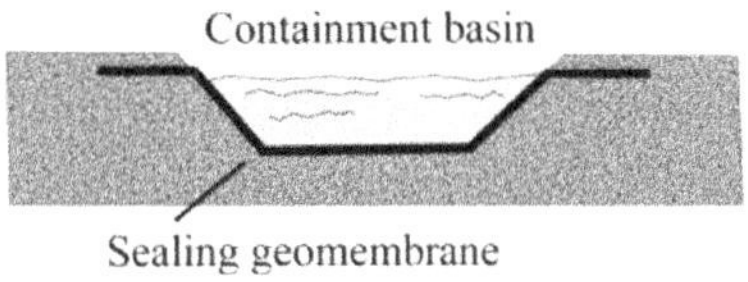

Figure 1.6. *Sealing geomembrane. For a color version of this figure, see www.iste.co.uk/villard/geosynthetics.zip*

1.2.6. *Protection*

Some geosynthetics can be easily damaged when directly exposed to static or dynamic forces from puncturing elements (granular material, scrap material, debris, etc.). To prevent perforation (Figure 1.7), it is essential to use a protection geosynthetic.

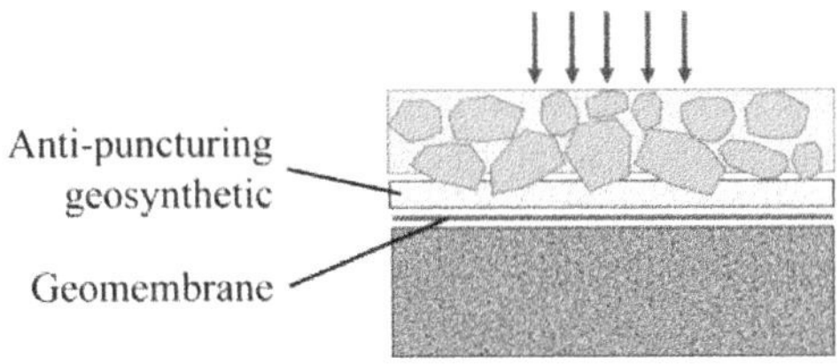

Figure 1.7. *Protection of a geomembrane against puncture by angular granulates. For a color version of this figure, see www.iste.co.uk/villard/geosynthetics.zip*

This material must have the adequate anti-puncturing mechanical characteristics to limit the damage to the underlying product.

1.2.7. *Erosion protection*

The erosion of slopes by rainwater or wind has a negative result: over time, could compromise the stability of the structures. In the case of the banks of canals and navigable waterways, repeated wave action is the cause of the scouring that can be observed at the level of the impact zone. To limit the movement and dislodging of granular particles on the surface, it is necessary to take erosion–protection measures (Figure 1.8).

For slopes and embankments subject to gully erosion or wind action, erosion–protection measures must protect surface soils long enough for hardy vegetation to be able to establish itself. Geosynthetics able to do this are of varying designs. They generally have an open mesh and are three-dimensional in nature in order to hold the soil in place and allow plant cover to develop. For banks subject to wave action, compact products, sometimes weighted down, can be used.

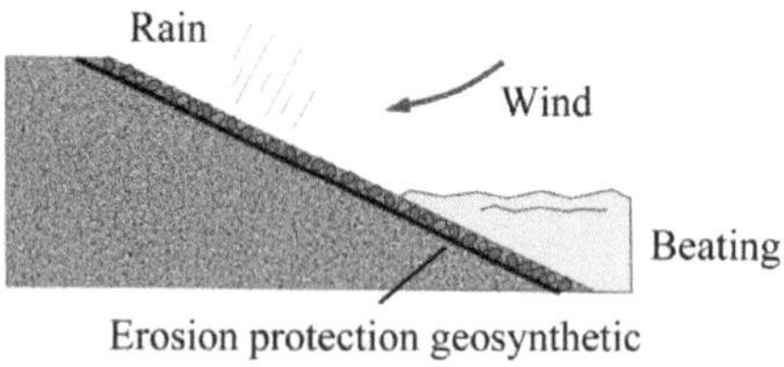

Figure 1.8. *Erosion protection geosynthetic: protection against rain, wind, and wave action. For a color version of this figure, see www.iste.co.uk/villard/geosynthetics.zip*

1.2.8. *Crack inhibition*

The use of a geotextile or geotextile-related product, sometimes impregnated with bitumen and integrated in a wearing course, makes it possible to slow down the development of reflective cracking (Figure 1.9). The goal of this technique is to extend the lifespan of roadways in order to minimize the number of maintenance and rehabilitation operations required. The CFG has drawn up "general recommendations for the use of geosynthetics as a means of slowing down the development of reflective cracking in roadways". It must be noted that in addition to the "stress relaxation" function and the creation of a barrier (sealing layer) that geosynthetics offer, it may be helpful to add a reinforcing component in the fight against reflection cracking in roadways. This is often achieved by incorporating glass fibers in the GTX being used. The crack-inhibition function is chiefly used in treating roadways and airport runways.

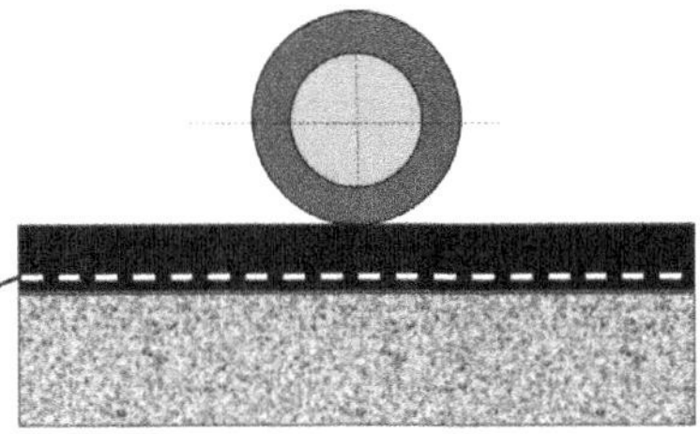

Figure 1.9. *Crack-inhibition geosynthetics for wearing courses. For a color version of this figure, see www.iste.co.uk/villard/geosynthetics.zip*

1.3. Properties of geosynthetics and principal characterization tests

1.3.1. *Mechanical properties*

The mechanical behavior of GTX depends on the nature of the base polymer, the manufacturing process of the reinforcing elements and the finishing process used. Mechanical properties can differ depending on the direction of production: machine direction (MD) or cross machine direction (CMD), especially for reinforcement geosynthetics.

1.3.1.1. *Tensile behavior of geosynthetics*

A knowledge of how geosynthetics behave under tension is essential when characterizing products. The mechanical characteristics of GTX and GTP are generally determined in tensile tests carried out on wide-width strips (the standard EN ISO 10319; BSI 2015b). For sealing geosynthetics, the tensile test differs depending on the nature of the product: GBR-P, GBR-B or GBR-C.

The test involves subjecting samples that are 20 cm wide and 10 cm long to tensile stress at a constant strain rate of 20% per minute. The load-strain curve (Figure 1.10) allows us to determine the tensile strength T, defined per meter of width (kN/m), the strain under maximum force ε_{Tmax} (%), the secant stiffness J (kN/m) calculated for a reference deformation ε_c and the quantity of energy absorbed (the area under the load-strain curve). The stiffness J of the geosynthetic affects both the deformability of the structure in which it is used and its behavior in use. The maximum tensile strength, T_{max}, and maximum strain, ε_{Tmax}, have a bearing on possible failure of the structure.

These characteristics may evolve over time, and this applies especially to creep deformation in the base materials, or to the results of chemical or mechanical alteration processes. This is especially important for structures reinforced using geosynthetics. To evaluate the creep deformation under tensile load, geosynthetic samples are subjected to a constant rate of loading (EN ISO 13431; BSI (2006b)). This allows for determination of the creep-rupture resistance as well as the evolution of strain under a constant force for a predefined service life of the structure (isochrone curves).

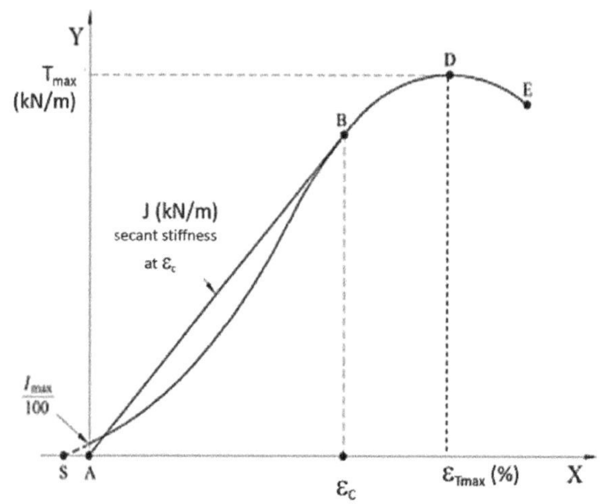

Figure 1.10. *Typical load-strain curve for a tensile test on a geotextile and geotextile-related product (EN ISO 10319; BSI 2015b)*

1.3.1.2. *Resistance to puncture and perforation*

When in use, geosynthetics may experience a drop in their physical and mechanical properties following local damage (which can extend up to perforation). These damages are the result of contact between the geosynthetic and angular elements (contact with granular soil, falling blocks, etc.).

Tests of puncture or perforation resistance aim to test the geosynthetic's ability to resist static or dynamic local forces. Depending on the type of test, the mechanical characteristics are as follows: puncture resistance (expressed in kN) or the dynamic perforation value (expressed in millimeters). These characteristics depend on the thickness, the mechanical properties of the base elements, the manufacturing method and the internal structure of the geosynthetic. The main standardized tests are as follows (Figure 1.11):

– The static puncture test (CBR): EN ISO 12236 (BSI 2006a). This test consists of determining the rupture force, in kN, required to push a flat-headed cylinder through the geosynthetic. The load is applied perpendicular to the sample at a constant velocity of 50 mm/min. This enables definition of the static puncture resistance of the product.

– Dynamic perforation test (cone-drop test): EN ISO 13433 (BSI 2006c). This test consists of dropping a conical steel object from a predefined height (0.5m) onto a geotextile held horizontally between two clamps. After the impact, the diameter (in millimeters) of the hole created is measured. The smaller the value obtained, the greater the geosynthetic's resistance to perforation. This test makes it possible to determine the product's resistance to dynamic perforation.

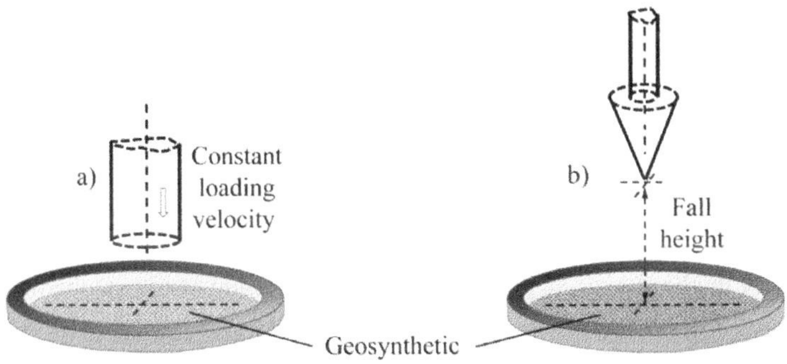

Figure 1.11. *(a) Static puncture tests and (b) dynamic perforation test. For a color version of this figure, see www.iste.co.uk/villard/geosynthetics.zip*

1.3.2. *Hydraulic properties*

1.3.2.1. *Permeability perpendicular to the plane and permittivity*

A geotextile's ability to allow water to pass through in a direction perpendicular to its plane is necessary to limit excess pressures in geotechnical structures. The behavior of the geotextile with respect to water flows is usually characterized by the coefficient of permeability normal to the plane, expressed in m/s (the ability to allow the passage of a fluid under the influence of a hydraulic gradient). The tests may be carried out without mechanical loading under a constant hydraulic load (standard EN ISO 11058; BSI 2019) or under a specific compression load (standard EN ISO 10776; BSI 2012). During the tests, the geosynthetic is subjected to a constant, unidirectional flow perpendicular to its plane (Figure 1.12). The permeability coefficient is defined as the ratio of the flow velocity (m/s) to the applied hydraulic gradient (m/m).

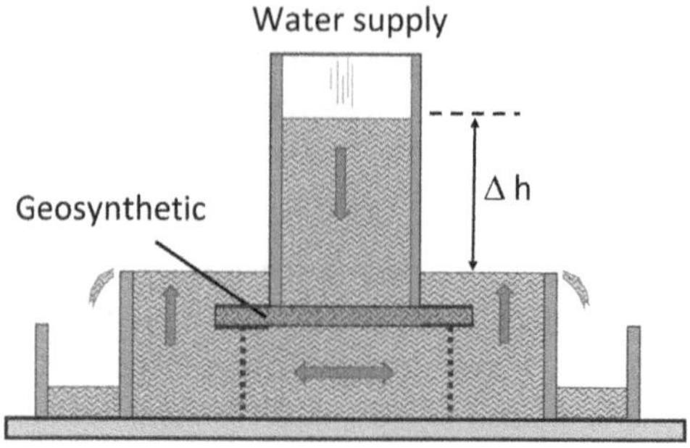

Figure 1.12. *Permeability perpendicular to the plane without mechanical stress. For a color version of this figure, see www.iste.co.uk/villard/geosynthetics.zip*

The permittivity ψ (standard EN ISO 10318-1; BSI 2015a) is the intrinsic ability of the geosynthetic to allow the passage of water perpendicular to its plane and corresponds to the ratio between the permeability coefficient and the thickness of the product. It is expressed per second. The volumetric flow rate through a geotextile, defined per unit area (in m/s), is therefore equal to the permittivity ψ (s^{-1}), multiplied by the head loss Δh (dissipation through friction of the mechanical energy of the fluid in movement, characterized by the height of an equivalent fluid column expressed in meters). The characteristic that is used in practice is the flow velocity for a head of 50 mm, VI_{H50} (m/s).

1.3.2.2. *In-plane flow and transmissivity*

A geotextile's ability to transport water in its plane is an essential element when planning and designing drainage projects. The geotextile's behavior with respect to water flows is usually characterized by the coefficient of permeability in the plane, expressed in m/s. Tests are carried out under constant hydraulic head (standards EN ISO 12958-1; BSI 2000a and EN ISO 12958-2; BSI 2020b). During the tests (Figure 1.13), the geosynthetic, under a fixed normal load, is subjected to a constant unidirectional flow in its plane. The coefficient of permeability in the plane is defined as the ratio of the flow velocity (m/s) to the applied hydraulic gradient (m/m).

The coefficient of transmissivity θ of a geosynthetic is a measure of its ability to transport water in its plane and thus allows for determination of the quantity of water that can be evacuated to an outflow. The unit of measurement is m^2/s. It is defined as the product of the coefficient of permeability in the plane and the thickness of the geosynthetic.

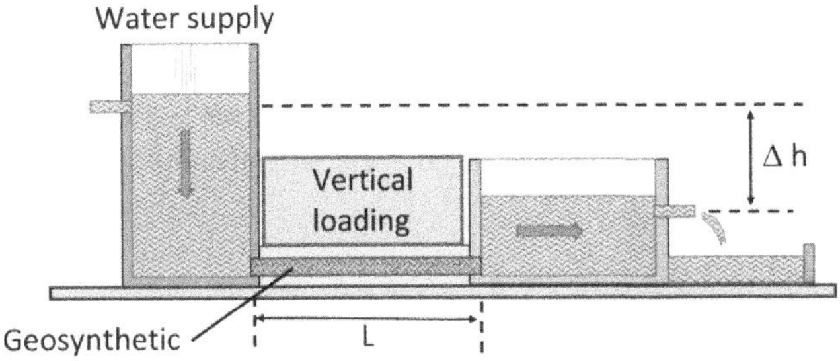

Figure 1.13. *Determining in-plane flow capacity. For a color version of this figure, see www.iste.co.uk/villard/geosynthetics.zip*

The volumetric flow of water transported in the plane of the geotextile per unit width of the sheet (m^2/s) is equal to $q = \theta\ \Delta h/L$, where θ is the transmissivity (m^2/s), Δh is the head loss (m) and L is the length of the geosynthetic sheet along which the liquid is flowing. The characteristic used in practice is the in-flow capacity under a stress σ, stated as $Q/l_{(\sigma)}$ (m^2/s).

1.3.2.3. *Characteristic opening size*

As mentioned in section 1.2.2, the challenge in designing filter geosynthetics is to optimize their opening size to allow for the movement of fluids while avoiding the migration of fine particles from one medium to another and preventing the clogging of the product.

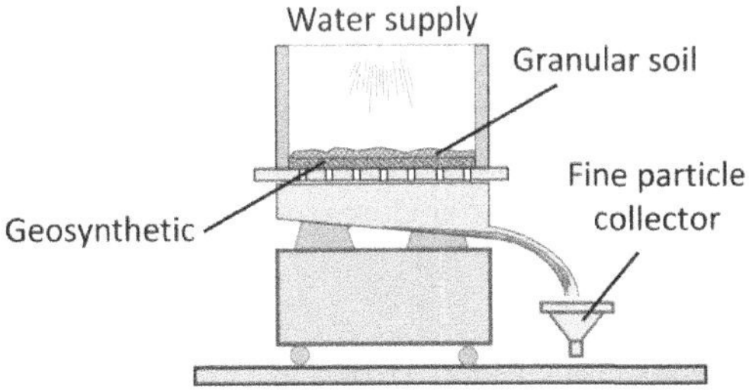

Figure 1.14. *Determining the characteristic opening size. For a color version of this figure, see www.iste.co.uk/villard/geosynthetics.zip*

The opening size O_f corresponds to the dimension (in μm) of the largest particle that can pass through the product. This is obtained (standard EN ISO 12956; BSI 2020c) by sieving a calibrated granular material washed through a sample of a geotextile acting as a sieve (Figure 1.14) and corresponds to the characteristic size of the material passing (that is, the value d_{90} of the particle-size curve of the material passing, which is the diameter of particles making up 90% of the material passing). The characteristic used in practice is the characteristic opening size $O_{90,\,W}$ (μm).

1.3.3. *Properties of the interface*

Interface friction is a complex phenomenon that allows shear forces to be transferred from one zone to another. It depends on the interface confining stress (normal stress on the interface), the size of the relative displacement between the two zones, the structural characteristics of the interface (texture and roughness) and certain complex mechanisms related, for example, to interlocking, entanglement or the rotation of particles in the case of an interface between a geosynthetic and soil.

While the principal mechanism of interaction for geosynthetic sheets is the friction with the soil, for open-mesh geogrids, there are several basic mechanisms that are involved: friction between the soil and the geogrid strands or ribs, the soil/soil friction within the openings in the grid and the blocking mechanism between soil particles and transverse ribs.

Characterizing the interface is a prerequisite for designing any structure that is a soil-geosynthetic composite. For low confining stresses, inclined plane testing of the interfaces is recommended (standard EN ISO 12957-2; BSI 2005). Under high confining stresses, depending on the desired loading method, suitable test methods use shear boxes (EN ISO 12957-1; BSI 2018h) and pull-out boxes (EN ISO 13738; BSI 2011). The interfaces tested are the soil-geosynthetic interface or the geosynthetic-geosynthetic interface, depending on the requirements of the application. In some cases, the characteristics of upper or lower interfaces must be determined when these have different textures, or if the soils in contact are different.

1.3.3.1. *Shear box test*

The shear box test (EN ISO 12957-1; BSI 2018h) is used for normal confining stresses greater than 5 kPa. The test set-up (Figure 1.15) is made up of two housings that are displaced with respect to each other at constant velocity. The geosynthetic sheet is generally attached to one of these two housings without any deformation being possible.

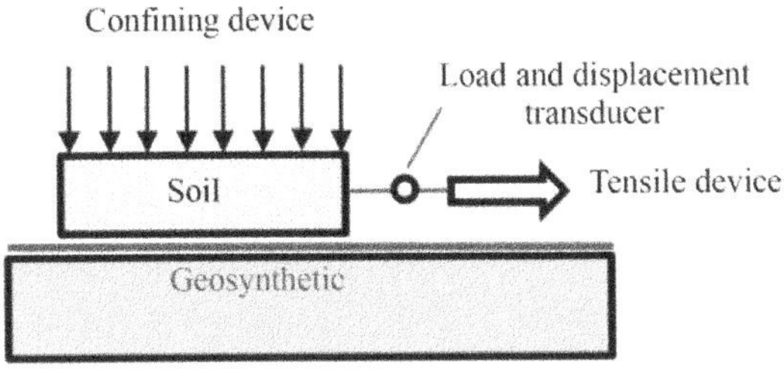

Figure 1.15. *Schematic depiction of the shear box test. For a color version of this figure, see www.iste.co.uk/villard/geosynthetics.zip*

When a coarse soil is tested, the size of the testing box must be appropriate for the maximum particle size of the soil being used. During the test, the movable housing is equipped with load and displacement transducers that allow the force–displacement curve for a given confining stress to be plotted. The shear stress for the interface is obtained by considering the tensile force and the area of the shearing surface.

Certain devices have a lower box that is longer than the upper box, so that the friction area remains constant with displacement, which is not the case when both boxes are identical in size. It must also be noted that the upper housing may lift slightly due to shearing of the interface. Several tests, with different confining stresses, are required to determine the friction angle δ and the adhesion a of the interface. These parameters depend on the state of density of the soil, which must be accurately determined.

1.3.3.2. *Inclined plane test*

With an applied level of confinement less than 5 kPa and the use of gravity loading, the inclined plane apparatus (EN ISO 12957-2; BSI 2005) appears best suited to characterizing the geosynthetic interfaces normally encountered in composite sealing systems on slopes (on embankments or cover layers) in waste storage facilities.

The apparatus (Figure 1.16) consists of a mobile housing that contains the soils to be tested, a mobile plane on which the geosynthetic is fixed and whose inclination is increased at constant speed and a lifting mechanism. The test set-up is equipped with a displacement sensor to determine the movement of the upper box, and in certain cases, it may also be equipped with a load sensor to determine the force required to prevent displacement of the upper box. The standard EN ISO 12957-2 (BSI 2005) recommends recording the friction angle of the soil-geosynthetic interface as the angle of inclination at which the housing slides with a displacement of at least 5 cm. But it has been shown (Briançon et al. 2011) that this procedure is not applicable to all interface types, and it is therefore preferable to determine the friction angle by measuring the force required to prevent displacement of the upper housing.

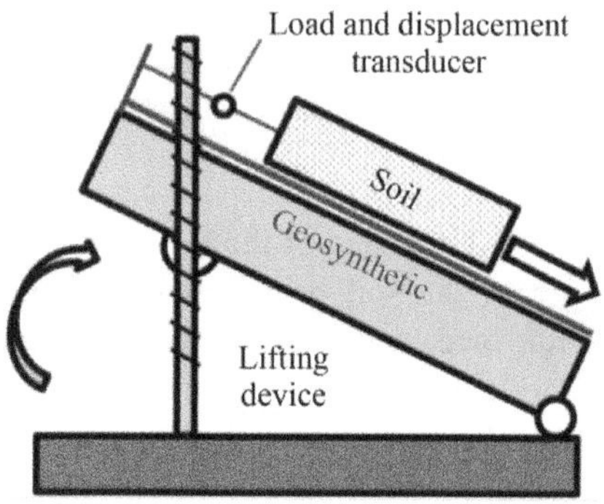

Figure 1.16. *Schematic depiction of the inclined plane friction test. For a color version of this figure, see www.iste.co.uk/villard/geosynthetics.zip*

As a result of gravity loading, the conventional test of sliding an object on an inclined plane yields information on the interface behavior up to the mobilization of maximum friction during the general sliding process. From a geotechnical point of view, the test conditions for the interface being tested are characterized by a free dilatancy of the soil and a normal stress that varies with the inclination.

The interface behaviors obtained from the inclined plane test differ from the behaviors that are usually observed with the direct shear box due to the differing test kinematics (Lalarakotoson et al. 1999). In the shear box test, the housing is displaced at a constant velocity, while in the inclined plane test, it is the angle of incline of the device and, consequently, the tangential force that is regulated. The inclined plane test also offers the advantage of demonstrating a jerky, "stick-slip" type of behavior, which is not possible with a shear box with a controlled rate of displacement.

On the other hand, it does not yield any information after the peak friction (unless an additional device is used) and does not therefore provide the residual friction value.

1.3.3.3. *Pull-out test*

The pull-out tensile test using a pull-out box (standard EN ISO 13738; BSI 2011) is used to determine the anchoring properties of a geosynthetic sheet confined within a soil (linear or trench anchoring). Under certain conditions (linear anchoring and taking a correction term into account), it can allow the friction angle at the interface to be determined.

The testing device (Figure 1.17) consists of a box containing the soil to be tested and a pull-out device. It is equipped with force and displacement sensors. The test may be carried out with or without a load on the upper layer.

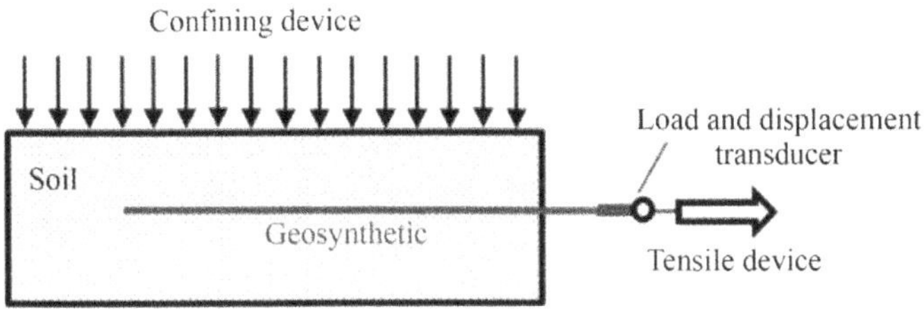

Figure 1.17. *Schematic depiction of the extraction box tensile grab test. For a color version of this figure, see www.iste.co.uk/villard/geosynthetics.zip*

The geosynthetic sheet anchored in the soil is pulled at a constant velocity out of the soil mass. It may stretch and undergo increasing deformations during the test. The maximum extraction force, T_{max} (kN), makes it possible to estimate the anchoring capacity T_A of the geosynthetic, expressed in kN/m (extraction force (T_{max}) divided by the width (d) of the geosynthetic test-piece).

For geosynthetic sheets, the anchoring capacity (T_A) per meter width of the sheet can be estimated ($T_A = 2\ f_b\ L\ \sigma \tan \delta$) by considering the interface friction δ, the length L of the test-piece, the confining stress at the interfaces (σ), with the two corresponding to the two soil-geosynthetic interfaces, and a correction term that depends on the apparatus and the experimental procedure being used (f_b).

1.4. Developments and additions

Following this general introduction and basic definitions, we recommend exploring the additional information and examples of these applications through the work of those who have worked with and implemented these applications throughout their careers. While not exhaustive, these examples illustrate the evolution of the design techniques and methods used in France for the past many decades. This volume concludes with a list of information gained from instrumented structures, which have been tracked over time and reanalyzed.

1.5. References

Briançon, L., Girard, H., Gourc, J.P. (2011). A new procedure for measuring geosynthetic friction with an inclined plane. *Geotextiles and Geomembranes*, 29(5), 472–482.

BSI (2005). EN ISO 12957-2, Geosynthetics – Determination of friction characteristics. Part 2: Inclined plane test. Standard, BSI, London.

BSI (2006a). EN ISO 12236, Geosynthetics static puncture test (CBR test). Standard, BSI, London.

BSI (2006b). EN ISO 13431, Geotextiles and geotextile-related products: Determination of tensile creep and creep rupture behaviour. Standard, BSI, London.

BSI (2006c). EN ISO 13433, Geosynthetics dynamic perforation test (cone drop test). Standard, BSI, London.

BSI (2011). EN ISO 13738, Geotextiles and geotextile-related products: Determination of pullout resistance in soil. Standard, BSI, London.

BSI (2012). EN ISO 10776, Geotextiles and geotextile related products: Determination of water permeability characteristics normal to the plane, under load. Standard, BSI, London.

BSI (2015a). EN ISO 10318-1, Geosynthetics – Part 1: Terms and definitions. Standard, BSI, London.

BSI (2015b). EN ISO 10319, Geosynthetics wide-width tensile test. Standard, BSI, London.

BSI (2016a). EN 13249, Geotextiles and geotextile related products: Characteristics required for use in the construction of roads and other trafficked areas (excluding railways and asphalt inclusion). Standard, BSI, London.

BSI (2016b). EN 13250, Geotextiles and geotextile related products: Characteristics required for use in the construction of railways. Standard, BSI, London.

BSI (2016c). EN 13251, Geotextiles and geotextile related products: Characteristics required for use in earthworks, foundations and retaining structures. Standard, BSI, London.

BSI (2016d). EN 13252, Geotextiles and geotextile related products: Characteristics required for use in drainage systems. Standard, BSI, London.

BSI (2016e). EN 13253, Geotextiles and geotextile related products: Characteristics required for use in erosion control works (coastal protection, bank revetments). Standard, BSI, London.

BSI (2016f). EN 13254, Geotextiles and geotextile related products: Characteristics required for the use in the construction of reservoirs and dams. Standard, BSI, London.

BSI (2016g). EN 13255, Geotextiles and geotextile related products: Characteristics required for use in the construction of canals. Standard, BSI, London.

BSI (2016h). EN 13256, Geotextiles and geotextile related products: Characteristics required for use in the construction of tunnels and underground structures. Standard, BSI, London.

BSI (2016i). EN 13257, Geotextiles and geotextile related products: Characteristics required for use in solid waste disposals. Standard, BSI, London.

BSI (2016j). EN 13265, Geotextiles and geotextile related products: Characteristics required for use in liquid waste containment projects. Standard, BSI, London.

BSI (2018a). EN 13361, Geosynthetic barriers – Characteristics required for use in the construction of reservoirs and dams. Standard, BSI, London.

BSI (2018b). EN 13362, Geosynthetic barriers – Characteristics required for use in the construction of canals. Standard, BSI, London.

BSI (2018c). EN 13491, Geosynthetic barriers – Characteristics required for use in the construction of tunnels and associated underground structures. Standard, BSI, London.

BSI (2018d). EN 13492, Geosynthetic barriers – Characteristics required for use in the construction of liquid waste disposal sites, transfer stations or secondary containment. Standard, BSI, London.

BSI (2018e). EN 13493, Geosynthetic barriers – Characteristics required for use in the construction of solid waste storage and disposal sites. Standard, BSI, London.

BSI (2018f). EN 15381, Geotextiles and geotextile-related products: Characteristics required for use in pavements and asphalt overlays. Standard, BSI, London.

BSI (2018g). EN 15382, Geosynthetic barriers – Characteristics required for use in transportation infrastructure. Standard, BSI, London.

BSI (2018h). EN ISO 12957-1, Geosynthetics – Determination of friction characteristics. Part 1: Direct shear test. Standard, BSI, London.

BSI (2019). EN ISO 11058, Geotextiles and geotextile-related products: Determination of water permeability characteristics normal to the plane, without load. Standard, BSI, London.

BSI (2020a). EN ISO 12958-1, Geotextiles and geotextile-related products: Determination of water flow capacity in their plane. Standard, BSI, London.

BSI (2020b). EN ISO 12958-2, Geotextiles and geotextile-related products: Determination of water flow capacity in their plane. Standard, BSI, London.

BSI (2020c). EN ISO 12956, Geotextiles and geotextile-related products: Determination of the characteristic opening size. Standard, BSI, London.

Lalarakotoson, S., Villard, P., Gourc, J.P. (1999). Shear strength characterization of geosynthetic interfaces on inclined planes. *Geotechnical Testing Journal*, 22(4), 284–291.

2

Reinforcement

Philippe DELMAS[1] and Jean-Pierre GOURC[2]

[1]*Formerly Cnam, Paris, France*

[2]*Formerly Polytech Grenoble IGE, Université Grenoble Alpes, France*

2.1. Introduction

The reinforcement of soil using sheets is a technique that has been under development since the earliest civilizations. As early as 2100 BCE, there were artificial constructions in Mesopotamia, ziggurats, which needed an enormous amount of materials and manpower, since they were reinforced by straw mats.

The first ziggurat, built in Uruk (Figure 2.1), was the model for all terraced structures that were built in Mesopotamia over the next thousand years or so. The best preserved ziggurats are in the city of Ur (built toward the end of the third millennium BCE).

Reinforcement techniques have evolved in many ways since then, including the most recent development of using geosynthetics for reinforcement. When arranged as a sheet, or as several superposed sheets, the higher mechanical strength of the material in one or more directions allows them to absorb tensile forces in the soil and limit deformation in extension. They thus make it possible to reinforce soils of low bearing capacity, prevent sink-hole formation above cavities and stabilize areas prone to slippage and landslides. They can, in effect, be used to create composite soil bodies in earthworks.

Harmonized European standards for geotextiles and geotextile-related products specify the applications for which the reinforcement functions can be used: roads

Geosynthetics from Yesterday to Today,
coordinated by Pascal VILLARD.

and other trafficked areas (BS EN 13249 (BSI 2016a), railways (BS EN 13250; BSI 2016b), earthworks, foundations and retaining structures (BS EN 13251; BSI 2016c), erosion-protection structures (BS EN 13253; BSI 2016d), reservoirs and dams (BS EN 13254; BSI 2016e), canals (BS EN 13255 (BSI 2016f)), solid waste disposals (BS EN 13257; BSI 2016g), liquid waste containment projects (BS EN 13265 (BSI 2016h)), and pavements and asphalt overlays (BS EN 15381 (BSI 2009)).

Figure 2.1. *Ziggurat constructed in Uruk. For a color version of this figure, see www.iste.co.uk/villard/geosynthetics.zip*

2.2. Soil reinforcement mechanisms

Soils that can be improved or reinforced are either the in situ soil present on the site or soil brought in from elsewhere, as is the case with reinforced embankments.

The majority of cases that involve soil improvement or reinforcement are listed in Figure 2.2 (according to NF P94-270 (AFNOR 2020a)). This list highlights the reinforcement of backfills with inclusions, including reinforcement using geosynthetics.

In the overwhelming majority of cases, the reinforcement geosynthetics are sheets or strips, placed horizontally between two fill layers. It is possible to differentiate between reinforcement using a single sheet of geosynthetic (sometimes doubled), as in embankments on soft soil, in reinforcements above cavities, or in the reinforcement of thin layers on a slope and reinforcements using multiple layers of geosynthetics in the soil body.

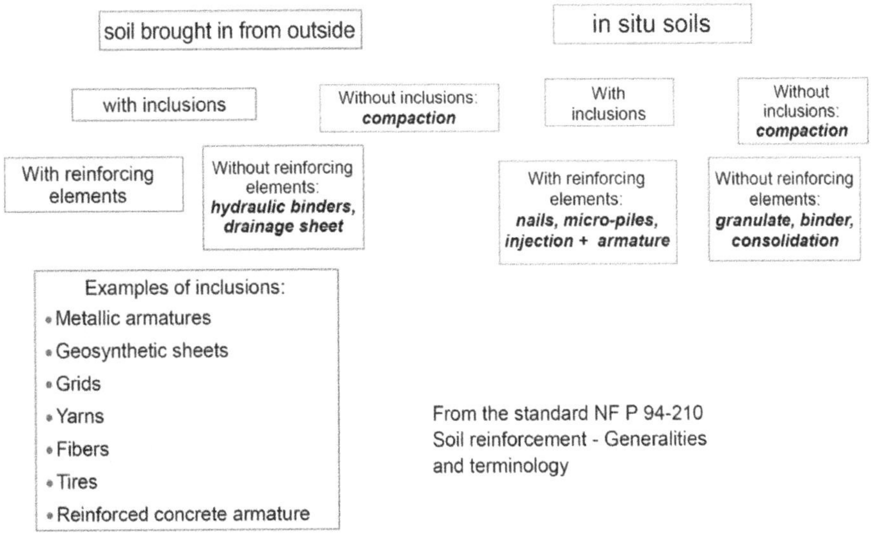

Figure 2.2. *Reinforcement and improvement of soils as per the standard NF P94-210 (AFNOR 1992). For a color version of this figure, see www.iste.co.uk/villard/geosynthetics.zip*

In the first case, the geosynthetic works in tension, like a reinforced-concrete steel bar, with an uptake of tensile forces through lateral anchoring (Figure 2.3).

In the second case, the soil-geosynthetic multilayer behaves as composite material. Therefore, when part of an embankment that is horizontally reinforced is subject to vertical loading (Figure 2.4), the uptake by the geosynthetic of the horizontal soil forces in the soil is equivalent to a lateral pressure on this part of soil, which can be seen as an increase in soil cohesion. It can then be said that the geosynthetics, aligned in the principal direction of soil strain, create soil confinement.

It is also possible to confine soil with reinforcement geosynthetics using bags, tubes or even containers (Figure 2.5).

However, this chapter will only be looking at reinforcement with horizontal sheets.

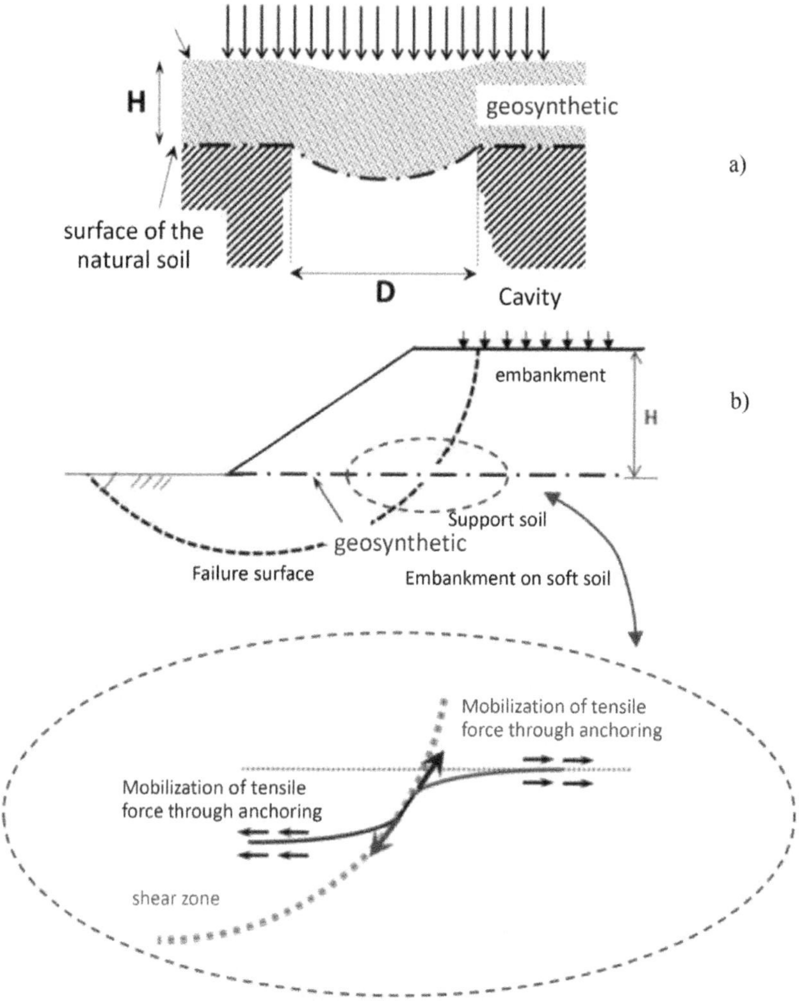

Figure 2.3. *Example of single-layer geosynthetic reinforcement with uptake of tensile forces by means of lateral anchoring(s): (a) embankment over a cavity; (b) embankment on soft soil. For a color version of this figure, see www.iste.co.uk/villard/geosynthetics.zip*

In conclusion, Figure 2.6 schematically summarizes the principal modes of deformation of a reinforcing element. With the exception of axial compression, geosynthetics can be subject to stress in all other modes: stress/strain, bending (also called membrane effect, for a geosynthetic) or extension/confinement (in the case of containers).

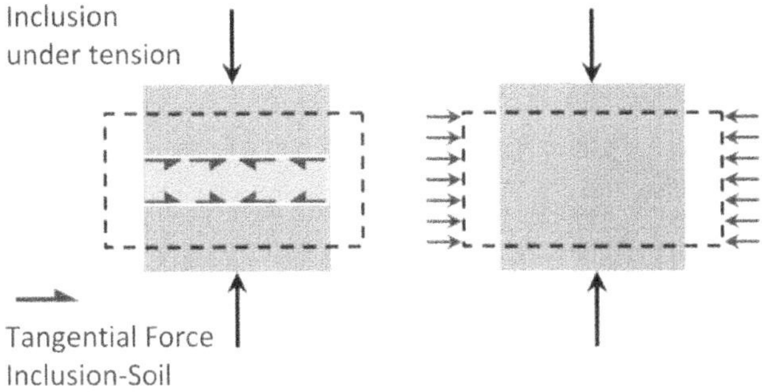

Figure 2.4. *Illustration of the principle of reinforcing a soil with a multilayer: the geosynthetic's uptake of horizontal soil forces (left) is equivalent to a lateral pressure on this body of soil (right). For a color version of this figure, see www.iste.co.uk/villard/geosynthetics.zip*

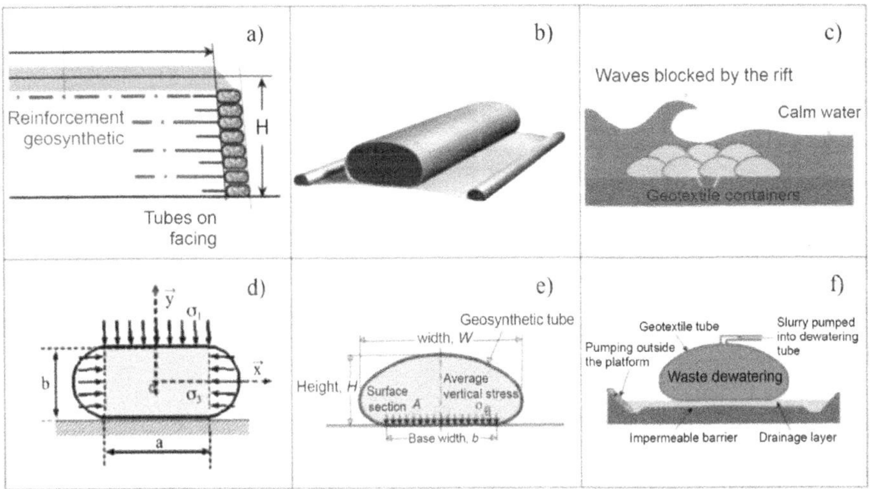

Figure 2.5. *Examples of the use of geosynthetic containers. For a color version of this figure, see www.iste.co.uk/villard/geosynthetics.zip*

COMMENT ON FIGURE 2.5.– *(a) Container-tube facing of a reinforced structure (Gorniak et al. 2013); (b) tubes for hydraulic applications (Lawson 2006); (c) marine applications (Lawson 2006); (d) mechanically filled and compacted containers (Gorniak* et al. *2013); (e) hydraulically filled containers (Lawson 2006); (f) containers for dewatering polluted sludge (Lawson 2006).*

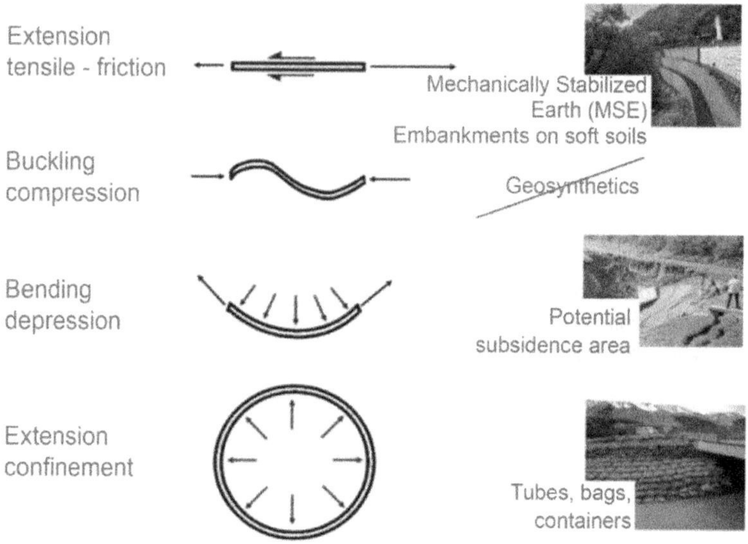

Figure 2.6. *Different modes of deformation of the reinforcing elements used to reinforce soils. For a color version of this figure, see www.iste.co.uk/villard/geosynthetics.zip*

2.3. Reinforcement geosynthetics and their characterization when used as reinforcement

2.3.1. *The different types*

In principle, it is possible to reinforce a structure using different types of geosynthetics, as shown by Baraize et al. (1995), who constructed and compared the reinforcement of surcharged soil bodies reinforced by non-woven geotextiles and geocomposites. However, in practice, reinforcement using geosynthetics is usually carried out using woven geotextiles, geogrids or geocomposites. The constituent polymers of these geosynthetics are generally, but not exclusively, polyester (PET), polypropylene (PP), polyethylene (PE), as well as polyvinyl acetate (PVA), or even aramid (AR).

2.3.2. *Nominal characteristics and general behaviors*

2.3.2.1. *Nominal characteristics*

The functional characteristics associated with reinforcement are defined in the standard BS EN 13251 (BSI 2016c) (Table 2.1).

It is possible to distinguish between the functional characteristics related to installation and those related to long-term behavior (durability) (Figure 2.7). These allow us to understand the characteristics used in the designing of geosynthetics over the lifespan of the structure in which they are used, as a function of the service requirements and the physicochemical evolution of the constituent materials of the geosynthetics.

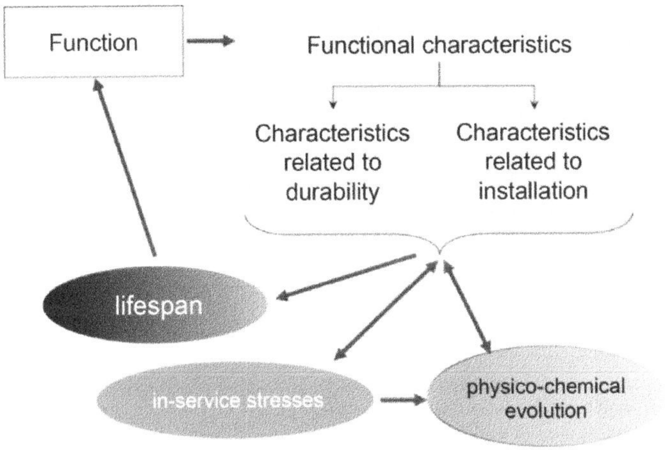

Figure 2.7. *Breaking down the properties of geosynthetics into functional characteristics linked to installation requirements and characteristics related to long-term behavior. For a color version of this figure, see www.iste.co.uk/villard/geosynthetics.zip*

Figure 2.8 depicts the typical evolution of a functional property as a function of the loading history of the geosynthetic from the time of installation until the end of service life.

In the case of reinforcement applications, the figure is especially applicable to the tensile strength of the geosynthetic. Starting from the initial strength of the product (Figure 2.8) as it leaves the production facility (A), the geosynthetic is subjected to a series of negative impacts related to handling (transport, storage), the installation on site and the compaction of fill layers, which results in a reduction in tensile strength (B). Subsequently, the tensile strength evolves over time depending on the service stresses, the environmental stresses and the type of geosynthetic (C), (D). With respect to the functional level of strength required by the design (R), the factor of safety is defined for the material at the end of the service life (RC) and at its real end of life (RD).

Characteristic	Test method	Function Reinforcement
(1) Tensile strength [b]	EN ISO 10319	A
(2) Elongation at maximum load	EN ISO 10319	A
(3) Tensile strength of seams and joints	EN ISO 10321	S
(4) Static puncture resistance (CBR test) [a b]	EN ISO 12236	A
(5) Dynamic perforation resistance (cone drop test) [a]	EN 918	A
(6) Friction (direct shear test)	prEN ISO 12957-1: 1997 and prEN ISO 12957-2:1997	A
(7) Tensile creep	EN ISO 13431	A
(8) Resistance to damage during installation under repeated loading	ENV ISO 10722-1	A
(9) Characteristic opening size	EN ISO 12956	–
(10) Water permeability normal to the plane	EN ISO 11058	A
(11) Durability	According to Annex B	A
(11.1) Resistance to climatic conditions	EN 12224	A
(11.2) Resistance to chemical ageing	ENV 12447, ENV ISO 12960 or ENV ISO 13438	S
(11.3) Resistance to microbiological damage	EN 12225	S
Relevance of codes: A: relevant to all conditions of use. S: relevant to specific conditions of use. "–": indicates that the characteristic is not relevant for that function.		
[a] Careful consideration should be given to the validity of these tests, which may not be applicable to certain types of product (e.g. geogrids). [b] If the mechanical properties (tensile strength and static puncture) are coded 'H' in this Table, this means that the producer must provide data for both properties. The use of only one, either tensile strength or static puncture, is usually sufficient in a project specification.		

Table 2.1. *Nominal characteristics associated with the reinforcement function defined in the standard BS EN 13251 (BSI 2016c)*

In certain applications where it is especially important to control deformations, for instance, in very high reinforced structures or in reinforcement over cavities, the above procedure is also adopted for the stiffness of the geosynthetics being used in order to limit the deformation of the structure at the end of its operational life.

Good design and an appropriate choice of geosynthetic ensure that the value of the functional property is greater than the operational level required for the operational life of the structure with an adequate safety factor.

In accordance with the harmonized European standards, the geosynthetics are characterized by nominal values (generally maximum values) and the ranges of relative variation (95% in general) obtained in standardized laboratory tests carried out on virgin samples from the factory.

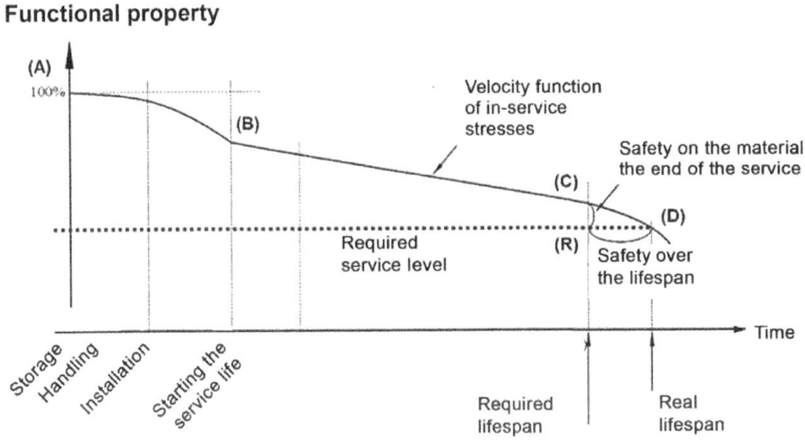

Figure 2.8. *Typical development of a functional property as a function of the history of stressing to which the geosynthetic has been subjected (ISO TS 13434 (ISO 2020))*

2.3.2.2. *Influencing factors (reduction factors)*

Considering the importance of the geosynthetic's deformation behavior in the behavior of the reinforced structures, it is important to understand the general principles of how a geosynthetic's behavior evolves over time.

There are three main factors that influence behavior over time:

– the development of the mechanical behavior toward damage during installation and compaction;

– the tensile creep behavior;

– the aging of the geosynthetic as a function of the environmental conditions.

– Mechanical behavior and resistance to damage: the evolution of the mechanical behavior of the geosynthetic during installation and compaction has been studied for a long time and has been integrated into all design standards governing reinforcement using geosynthetics (e.g. NF G38-064 (AFNOR 2016)) in the form of a reduction factor for tensile strength. Appendix C of this standard specifies the various methods of determining the reduction in strength in accordance with the stress levels to which the product is subjected. Apart from strength, the mechanical behavior of geosynthetics is also affected by a possible variation in stiffness during installation and compaction. This phenomenon has been studied by many authors. Müller-Rochholz (2003), in particular, differentiated between external damage, which corresponds to cuts in the reinforcing components of the geosynthetic, and internal damage that the reinforcing elements may sustain when they are covered by a protective coating or another geosynthetic. In this case, the integrity of the reinforcing components may be relatively well preserved. Figure 2.9 shows that while geosynthetics that suffer external damage exhibit a reduction in stiffness as a result of the installation and compacting processes, for geosynthetics that only suffer internal damage, there is no loss of stiffness.

– Creep behavior: the impact of tensile creep on tensile strength in geosynthetics has been studied for a long time and has been integrated into design standards for reinforcement using geosynthetics (e.g. NF G38-064 (AFNOR 2016)) in the form of a reduction factor. The impact of creep on deformation is often represented using isochrone curves (BS EN ISO 13431 (BSI 1999)) (Figure 2.10).

– Resistance to chemical aging: various authors have tried to establish laws relating to behavior over time for the aging process (Elias et al. 1988; Nait-Ali and Freitag 2014; Nait-Ali et al. 2014).

The loss of tensile strength depends on the type of degradation and can be addressed using specific models. This decrease in strength is integrated into all the design standards for reinforcement using geosynthetics (e.g. NF G38-064 (AFNOR 2016)) in the form of a reduction factor.

However, the impact of creep on stiffness has not been studied in much detail. Nonetheless, this can be stated: as far as hydrolysis is concerned (in neutral or acidic pH), in general there is no reduction in stiffness (or a very small reduction), which is not the case with alkaline hydrolysis, where there is a noticeable loss of stiffness from the beginning. As far as other types of degradation, such as oxidation, are concerned, the reduction in mechanical properties is different. In certain cases, a noticeable increase in stiffness is observed, related to an increase in the degree of crystallinity.

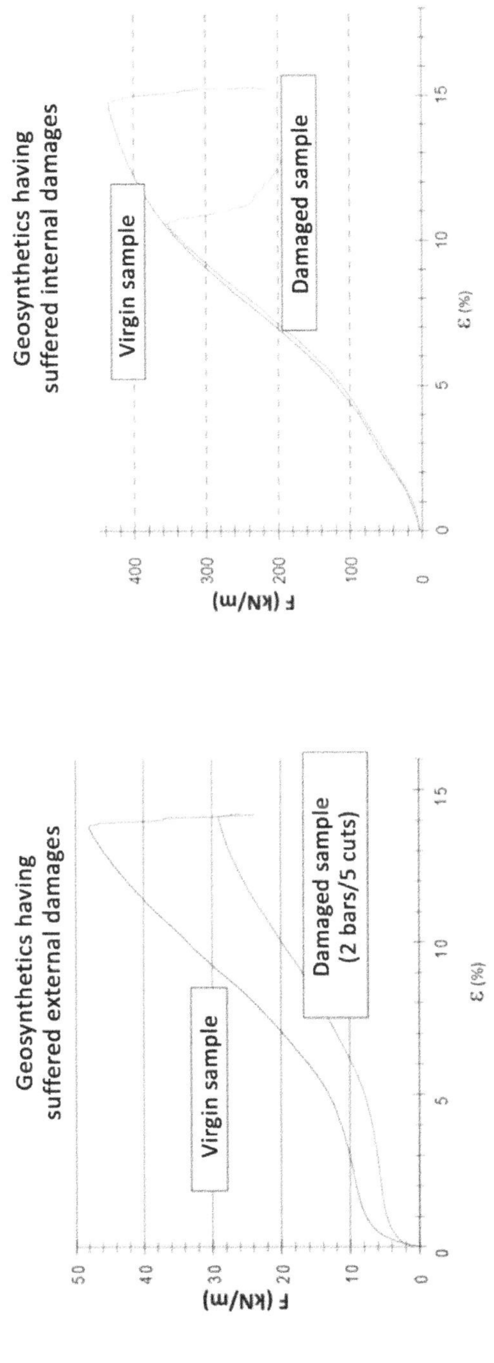

Figure 2.9. *Impact of installation and compaction on stiffness due to different types of damage. For a color version of this figure, see www.iste.co.uk/villard/geosynthetics.zip*

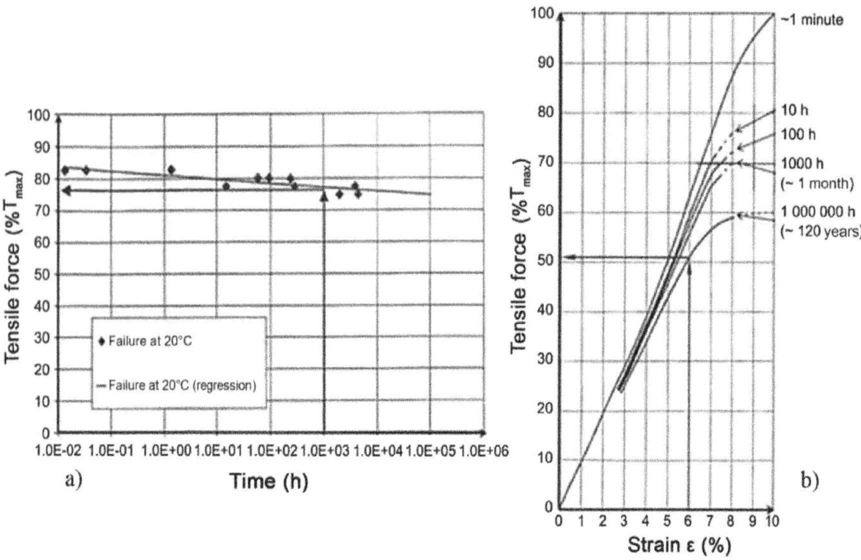

Figure 2.10. *Impact of creep on tensile strength e (a) and on the deformation behavior (b)*

2.3.2.3. *Permissible long-term characteristics*

When the nominal characteristics for the functions that the structure must perform are known, it is possible to calculate the permissible long-term characteristics of the geosynthetics, that is, for the working life of the structure. Taking the example of the characteristic of the tensile strength of the geosynthetic, it is possible to evaluate the corresponding nominal value, called "short-term" characteristic strength ($R_{t;k}$), measured in accordance with the standard NF EN ISO 10319 (AFNOR 2015). The ultimate tensile strength of the geosynthetic $R_{t;d}$ is defined in the French standards (NF G38-064 (AFNOR 2016), NF G38-067 (AFNOR 2017) and XP G38-065 (AFNOR 2020)):

$$R_{t;d} \leq (1/\Gamma_{geo}) R_{t;k} \qquad [2.1]$$

where Γ_{geo} is the factor which contains:

– the partial factor of tensile strength ($\gamma_{M;t}$) defined in the Eurocode 7 standard (BS EN 1977-1 (BSI 2013a));

– the reduction factors that enable the geosynthetic's behavior over time to be taken into account:

$$\Gamma_{geo} = \gamma_{M;t}.\Gamma_{CR}{}^{*}.\Gamma_{CH}{}^{*}.\Gamma_{ID*} \qquad [2.2]$$

where:

– $\gamma_{M;t}$ is the partial factor for tensile strength;

– $\Gamma_{CR}{}^*$ is the reduction factor linked to the reduction in tensile strength as a result of creep in the geosynthetic reinforcement;

– $\Gamma_{CH}{}^*$ is the reduction factor related to the reduction in tensile strength due to the aging of the geosynthetics related to environmental conditions;

– $\Gamma_{ID}{}^*$ is the reduction factor corresponding to the reduction in tensile strength due to damage sustained by the geosynthetic reinforcement during installation and the compaction of the embankments.

NOTE.– In this chapter, the reduction factors have been modified from the French version of the standards for the English speaking readers:

– Γ_{instal}, Γ_{flu} and Γ_{vieil} have been replaced, respectively, by $\Gamma_{ID}{}^*$, $\Gamma_{CR}{}^*$ and $\Gamma_{CH}{}^*$.

– Similarly, the reduction factors applied on the force of the geosynthetic at the in-service strain (ε) introduced in the French standards, named C_{instal}, C_{flu} and C_{vieil}, have been replaced, respectively, by $C_{ID}(\varepsilon)^*$, $C_{CR}(\varepsilon)^*$ and $C_{CH}(\varepsilon)^*$.

2.3.2.4. *Characterizing the tensile behavior of the reinforcement*

As previously discussed (section 1.3.1.1), the geosynthetic's creep behavior is an essential element when designing reinforced structures. This behavior is generally characterized in a tensile strength test by three parameters (Figure 1.10): the tensile strength T_{max} (kN/m), the deformation at maximum load ε_{Tmax} (%) and the secant stiffness $J_{\varepsilon c}$ (kN/m) defined for a reference deformation ε_c. The manufacturing method used to produce the geosynthetic – weaving, knitting, extrusion and drawing, etc., of the constituent yarns or wires – as well as the choice of polymer, result in different tensile behavior.

Along with the most common geosynthetics, which take the form of sheets, there are a few geosynthetic-reinforcement technologies in the form of strips, and frequently also cables made up of coated yarns. In this case, the tensile strength is reported per meter of length of the structure as a function of the spacing of the strips.

The BS EN ISO 10321 (BSI 2009) standard defines the method for measuring tensile strength of seams and joints. However, seams are rarely used in reinforced structures as their strength is generally lower than the product's nominal strength. Every attempt is therefore made to avoid joints and seams in the direction of reinforcement.

It must be noted that for the design of long-term structures, information on the short-term tensile behavior of the geosynthetic must be complemented by information on creep, aging and resistance to damage (section 2.3.2.2).

The BS EN ISO 13431 (BSI 1999) standard is used to evaluate the creep behavior of the products. The samples are placed under stress at a constant value and the test determines the tensile creep behavior and the creep rupture of the geosynthetic. Given that the test is carried out over a long period and is relatively difficult to conduct, this cannot be considered a routine quality-control test. For faster results, the Stepped Isothermal Method (SIM) is often used. This involves testing different geosynthetic samples over a short period at different temperatures and using the time–temperature correlation to establish the long-term creep curve.

The creep test allows two important parameters to be determined:

– the loss of tensile strength related to creep, characterized by the reduction factor (Γ_{CR}*);

– the deformation of the product under tensile stress over time, characterized by the isochrone curves that relate the couples (force, deformation) obtained during the creep test over the same time (Figure 2.10).

In the absence of data on creep for the geosynthetic, the French design standards (e.g. NF G38-064 (AFNOR 2016)) propose the values given in Table 2.2 for the reduction factor (Γ_{CR}*), corresponding to the loss in tensile strength in the product over time based on the polymer used for reinforcement.

Long-term structure (120 years)	PP/PE	PEHD	PET/PVA/AR
Γ_{CR}*	6	5	3

Table 2.2. *Default values of the creep reduction factor (Γ_{CR}*) (NF G38-064 (AFNOR 2016))*

The resistance to damage during installation of reinforcement geosynthetics should be evaluated by the use of the European standard BS EN ISO 10722 "Resistance to damage during installation" (Row 9 in Table 2.1). Nevertheless, as this laboratory test is deemed to be non-representative of a real-world situation, it is not used, and it is advised to use the results obtained with the full-scale test as far as possible (BS EN 17738 (BSI 2023)).

The French standards recommend that the resistance to damage during installation of reinforcement geosynthetics should be determined based on full-scale tests with different soils and at different levels of compaction. This may be done to determine the product's loss of resistance for the four typical conditions, ranging from "not very severe" to "extremely severe" (Table 2.3). In the absence of experimental data, the values given in Table 2.3 for the reduction factor (Γ_{ID}*) corresponding to the product's loss of strength during installation and the compaction of the underlying soil are used.

The resistance to chemical aging of a geosynthetic is determined in accordance with the procedure defined in Appendix B of the BS EN 13251 (BSI 2016c) standard. In the absence of data, the French design standards propose the values given in Table 2.5 for the reduction factor (Γ_{CH}*) corresponding to the loss of strength of the product for service lives of 50 years and 100 years, as a function of the reinforcing polymer and on the soil pH.

	$D_{max} \leq 50$ mm			$D_{max} > 50$ mm
Material	Fine soils Sands	Gravelly soils Sandy soils		Materials with coarse fractions
	–	Alluvial	Crushed	Rounded or angular
Medium compaction energy	Not very severe	Average severity	Severe	Extremely severe
Intense compaction energy	Average severity	Severe	Extremely severe	Not recommended

Table 2.3. *Degree of severity of the installation conditions for a layer with a thickness of 30 cm*

	Installation conditions			
All geosynthetics	Not very severe	Average severity	Severe	Extremely severe
	1.10	1.25	1.50	2.50

Table 2.4. *Values for the default reduction factors (Γ_{ID}*) corresponding to the loss of product strength during installation and the compaction of the overlying soil*

PP	HDPE (extruded)	PET (fibers)	PA	pH	Service life
1.3	1.3	1.2	–	4–8	100 years
1.3	1.3	1.3	–	8–9	
1.05	1.05	1.05	1.0	4–8	50 years
1.05	1.05	1.1	1.0	8–9	

Table 2.5. *Default values of the reduction coefficient (Γ_{CH}*) as function of the reinforcing polymer and the soil pH*

2.3.2.5. *Characterizing the membrane behavior of the reinforcement*

As depicted in Figure 2.11, when the geosynthetic is subject to normal loads, these may be withstood in tension thanks to what is called the "membrane effect". For instance, this is what happens when a geotextile used as reinforcement over zones that are susceptible to collapse absorbs loads from the structure overlying a cavity. The tensile forces that develop in the geotextile must be withstood laterally, on either side of the cavity, by means of appropriate anchoring systems. There are no standardized tests to determine the behavior of a geosynthetic as a membrane depending on its tensile behavior.

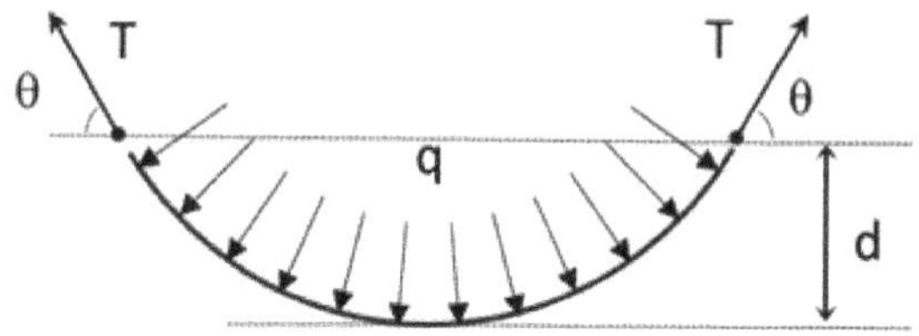

Figure 2.11. *Principle of the membrane effect*

Figure 2.12 shows (1) the increase in settlement (d) as a function of the normal stress (q) as well as (2) the tensile force (T) for the case of a sheet with fixed ends and a stiffness J = 5,000 kN/m. Similarly, for a normal stress q = 20 kPa, the curve (3) shows the influence of stiffness (J) on the settlement (d) and (4) the tensile force (T). It can be seen that the settlement (d) can be controlled, or even reduced, by increasing stiffness (J). On the other hand, this results in a large increase in the tensile force (T).

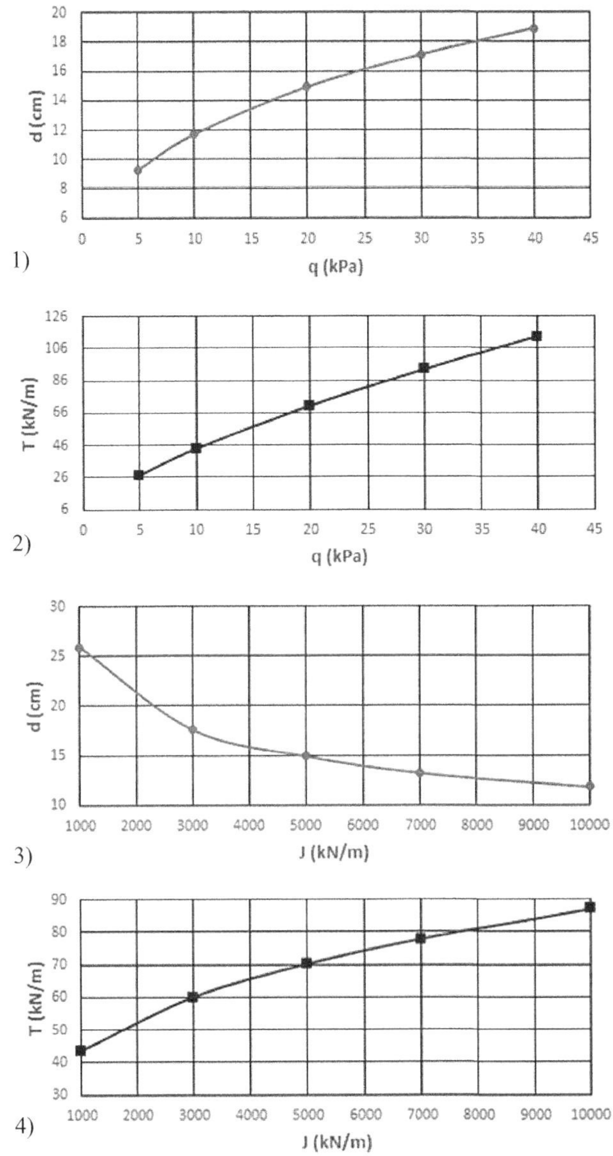

Figure 2.12. *Demonstrating the membrane effect: (1) settlement d and (2) tensile force T as a function of the normal load q for a stiffness J = 5,000 kN/m; (3) settlement d and (4) the tensile force T as a function of the stiffness J for a normal load q = 20 kPa*

2.3.2.6. *Mobilizing the anchoring forces of reinforcement geosynthetics*

The mobilization of anchoring forces results from the coupling of the geosynthetic's tensile behavior and its friction behavior in contact with soil as a function of the relative soil-geosynthetic displacement (section 1.3.3).

If it is considered that the deformation of the geosynthetic remains less than 5%, it is often reasonable to consider that the geosynthetic's behavior can be described by a linear elastic law, as well as an elastoplastic law for the interface behavior, as many laboratory studies have demonstrated. The half-scale test carried out by Bordeau et al. (1990, 1994) illustrated the progressive mobilization of a geosynthetic sheet under tension. The sheet, 1.5 m long, is buried to a depth H = 1.05 m in a soil mass (test in a parallelepipedic tank). Displacements along the sheet are measured during the pull-out test at different points located at a distance x from the free end of the sheet.

T_A is the anchoring force at the head (point A – x = 1.5 m), measured as a function of the displacement of the head of the anchored sheet u_A. The schematic diagram (Figure 2.13) depicts the displacements along the sheet during the pull-out test. Point B (x = 0) corresponds to the free end of the sheet.

It can be seen that the length of the sheet slipping relative to the soil (which mobilizes the interface friction) increases progressively along with the anchoring force T_A: the point located at x = 0.75 m only begins to slip for a head anchor displacement T_A = 25 mm, and the free end of the sheet, B, only begins to move at a displacement u_A = 55 mm.

The total length of the sheet is thus mobilized to slip with respect to the soil and only a small additional displacement is required to develop the maximum anchoring force T_A (u_A = 73 mm).

The stiffness, J, of the geosynthetic influences the behavior of the anchored sheet, as demonstrated in the centrifuge tests carried out at the LCPC in Nantes under an acceleration of 30 g (Gourc et al. 1992).

The displacement required to mobilize anchoring forces reduces as the stiffness of the sheet increases (example of the comparison between a geotextile and a metallic sheet (Figure 2.14)).

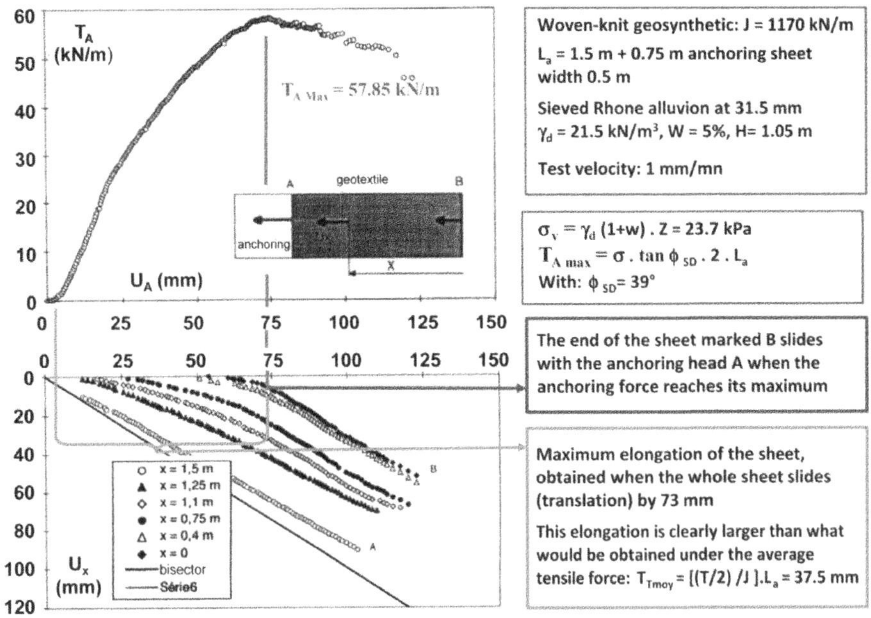

Figure 2.13. *Schematic diagram depicting displacements along the sheet over the course of the pull-out test. For a color version of this figure, see www.iste.co.uk/villard/geosynthetics.zip*

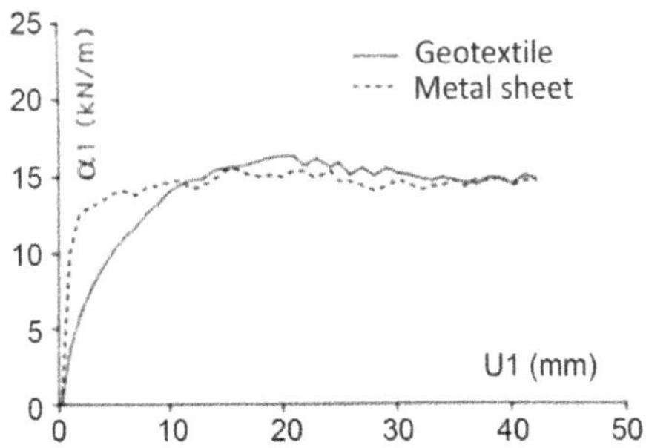

Figure 2.14. *Comparison of the anchoring forces mobilized in a centrifuge with an acceleration of 30 g (a) as a function of the displacement of the anchor head (U1) for a geotextile (stiffness of the prototype, J_g = 106 kN/m) and a metallic sheet with a thickness of 1 mm (stiffness, J_m = 210,000 kN/m) (Gourc et al. 1992)*

It is also possible to calculate, for a flat anchorage, the importance of the stiffness of the geosynthetic in mobilizing the anchoring force. For example, Figure 2.15 demonstrates (for an elastoplastic interfacial behavioral law) that for the same tensile force (T = 18.9 kN/m), a geosynthetic with a relatively small stiffness (J = 500 kN/m) mobilizes friction forces over an anchorage length of 2 m, while for a geosynthetic with a greater stiffness (J = 2,000 kN/m), the friction forces are mobilized over a length greater than 4.5 m.

This is related to the relative displacements of geotextiles to soil that differ according to the rigidities of geosynthetics. A stiffer geosynthetic mobilises less friction locally, but along a longer length, where as a more extensible geosynthetic mobilises more friction locally (especially close to the head of the anchor), but along a shorter length. This result shows that, if sufficiently distant from the maximum pull-out force, the mobilization lengths of the operational forces are smaller for “extensible” reinforcements such as geosynthetics than for reinforcements with “low extension”, such as metallic reinforcements. This may be considered to be a definite advantage if we also verify that the corresponding deformations in the geosynthetics, and therefore the structure, remain permissible.

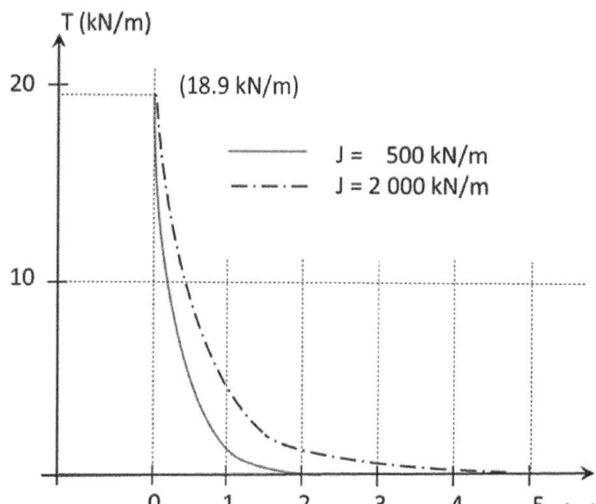

Figure 2.15. *Influence of the stiffness of the geosynthetics on the mobilization of tensile forces along an anchorage (Delmas et al. 1986a)*

Certain applications, such as the reinforcement of thin layers on a slope, or an embankment over cavities, require the use of trench anchors. Although the corresponding design norms use very simplified rupture-calculation methods to assess the maximum forces that can be mobilized in a trench anchor, Villard and Chareyre (2004) propose taking into account how the shape of the geosynthetic (using the angle of curvature at the top and bottom of the trench: pulley effect) affects the force that can be mobilized (Figure 2.16).

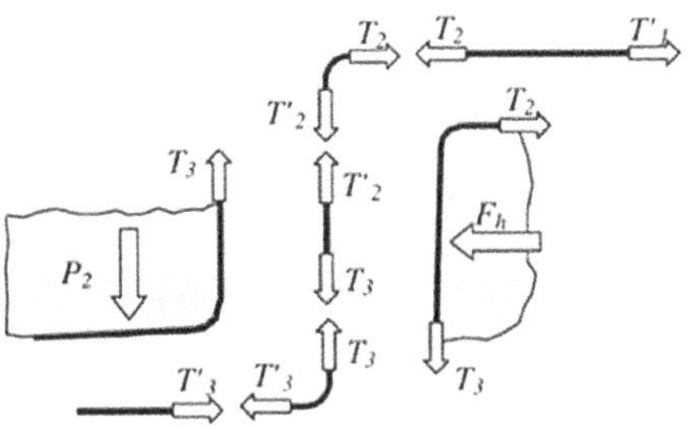

with:

$$T_3' = 2P_2 \tan(\delta)/[1 + 2K_3 \tan(\delta)]$$

$$T_3 = 2P_2K_3 \tan(\delta)/[1 + 2K_3 \tan(\delta)]$$

$$K_3 = \exp(\pi/2 \tan\delta)$$

δ soil-geosynthetic friction angle

Figure 2.16. *Taking into account the shape of the geosynthetic, making use of the effect of an angle at the top and bottom of the trench on the force that can be mobilized*

2.4. Design and installation of geosynthetic reinforcements

There are design standards for the main types of structures reinforced by geosynthetic (Table 2.6). All these standards follow Eurocode-compliant design principles ((BS EN 1990 (BSI 2007), BS EN 1977-1 (BSI 2013a) and BS EN 1977-2 (BSI 2013b)), where the definition of the important categories and site conditions is followed by the selection of potential unfavorable locations for the project, and the limit states to be considered: verification of the ultimate limit states (ULS) to confirm the stability of a reinforced structure during construction and operation, and the verification of in-service limit states to prove that the deformations of the reinforced structure are acceptable during construction and operation.

The general principle behind these design standards consists of carrying out a stability calculation in accordance with the Eurocode principles (ULS) by weighting the actions (permanent or variable, favorable or unfavorable) and the soil parameters (strength, density), by applying the factors associated with the geotechnical calculation (bearing capacity, sliding, subsoil strength and model partial factor).

Type of reinforcement structure	Standards
Inclined walls and slopes reinforced by geosynthetic sheets	NF G38-064 (1) (AFNOR 2016)
Retaining structures – reinforced embankments and soil nailing structures	NF P94-270 (2) (AFNOR 2020)
Basal reinforcement of embankment prone to soil subsidence with geosynthetics	XP G38-065 (AFNOR 2020)
Stabilization of a thin layer on a slope	NF G38-067 (AFNOR 2017)
Execution of special geotechnical projects – reinforced embankments	BS EN 14475 (BSI 2006)

(1) Vertical wall and battered wall ≥ 1H/4V.

(2) Reinforced, inclined wall or embankment and reinforced slope: batter < 1H/4V.

Table 2.6. *Standards validating the design and design elements of structures reinforced using geosynthetics, and standards for execution*

The ultimate tensile strength $R_{t;d.}$ of the geosynthetic is used in the stability calculation as a stabilizing element. It is calculated from the short-term characteristic tensile strength ($R_{t;k}$) of the geosynthetic, weighted by the partial factors of the tensile strength and the reduction factors for the long-term behavior of the product (equation [2.2]).

On the other hand, in certain structures which are particularly sensitive to deformation, the standards recommend a Eurocode-compliant calculation (serviceability limit states (SLS)). In this case, the deformations in the structure are evaluated to ensure that they are compatible with a normal functioning of the structure. The standard XP G38-065 (AFNOR 2020) for reinforcement above potential soil subsidence areas recommends carrying out a calculation for the deformations in the geosynthetic and the settlement on the surface of the embankment. This is based on simple constitutive laws for the soil-geosynthetic interaction and for the tensile behavior of the geosynthetic. On the other hand, for retaining structures, limits on the maximum deformation of the geosynthetic are fixed according to the sensitivity of the structure (NF G38-064 (AFNOR 2016) and NF P94-270 (AFNOR 2020)).

In general, we can refer to existing norms for planning and designing the vast majority of structures reinforced using geosynthetics.

Consequently, the rest of this chapter will introduce examples of the planning of actual structures and their design and, wherever necessary, introduce the relevant installation method. This is done to try and expand upon the cases covered by the standards and to demonstrate a few specific approaches that can be used for designing real structures.

2.4.1. *Reinforcing an embankment on soft soil: a case of deliberately accepted partial punching (Aulnois embankment)*

2.4.1.1. *Problem statement for the structure*

The motorway near Aulnois (France) needed to cross a site containing waste from a beet-washing facility consisting of 6 m of highly compressible organic residue overlying a peat layer 3 m thick. This planning involved constructing the embankment in three phases, knowing that the final surface of the project would be just 80 cm above the organic waste (Figure 2.17).

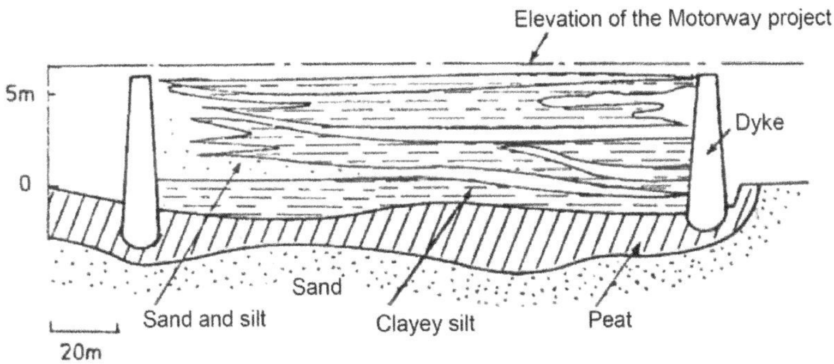

Figure 2.17. *Geological profile of the project*

The geotechnical characteristics of the soils in the centre of the basin are given in Table 2.7. It can be seen that the residue in the basin is highly heterogeneous. This is partly linked to the process of constructing a dike every year by recovering the sandy components deposited during the preceding washing campaign.

On the other hand, the organic matter in the deposit, as well as the underlying layer of peat indicate there will be a secondary, long-term consolidation of soils under the weight of the motorway embankment, and this must be taken into account when planning the solution and designing the geosynthetics.

	γ (kN/m^3)	w (%)	e_0	σ'_c (kPa)	Cs	Cc	c_u (kPa)
Residue (0.6)	12.9 to 18.8 (16)	15.5 to 101	0.745 to 2.29 (1.72)	21 to 48 (31)	0.01 to 0.035 (0.032)	0.632 to 0.841 (0.632)	7 to 19 (14)
Peat (6.9 m)	9.4 to 10.9	245 to 440	3.11 to 6.6 (4.82)	29 to 46 (46)	0.065 to 0.133 (0.065)	3.3 to 4.74 (3.3)	28 to 36 (30)

Table 2.7. *Geotechnical characteristics of the soil at the centre of the basin*

2.4.1.2. *Proposed solution: planning and design of the elements*

The chosen solution was a deliberately accepted partial punching followed by decreasing the weight of the embankment (Delmas et al. 1986b). The phases are depicted in Figure 2.18.

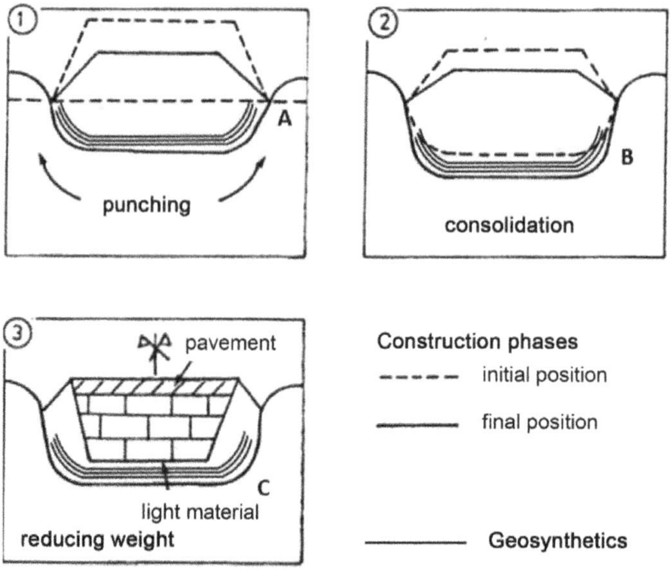

Figure 2.18. *Phases of the solution chosen*

This comprised:

– deliberately accepted partial punching of the residue with the aim of reducing its depth, since it was both compressible and susceptible to creep under the embankment;

– loading for a period long enough to allow partial pre-consolidation of the remaining sub-soil;

– unloading of the embankment to a height that allowed the construction of a light embankment that could support the final road. This final phase allowed the artificial "aging" of the compressible soil from the point of view of secondary consolidation (Bjerrum 1967). Figure 2.19 depicts (e, log σ'_v) the principle underlying the path taken by the effective vertical stress at a point in the residue located 0.6 m below the embankment during the construction phase, in accordance with Bjerrum (1967).

The initial state of the residue (just after the partial punching) is given by point A. The consolidation under the loading by a 5 m embankment leads to the effective stress at point B, and the unloading and the construction of the final road to the effective stress on C. Finally, the effective long-term stress is represented by point D.

The "classical" construction of the final embankment and the roadway would have been represented by point B' at the end of construction, and by D in the long term.

It can be seen that the proposed phases (A, B, C), carried out over a short period of about six months lead to the effective stress shifting from point A to point C (Figure 2.19), whereas conventional construction (A, B', C) would take several decades. During this period, there would be significant secondary settlement, resulting in significant maintenance costs for the road.

The design of this solution, which is always tricky anyway, is complicated here by the absence of representative data on the creep characteristics of the residues and the uncertainty related to the primary consolidation parameters associated with the heterogeneity of the material. For all these reasons, the designing team decided to first construct an experimental embankment in order to finalize the design of the structure.

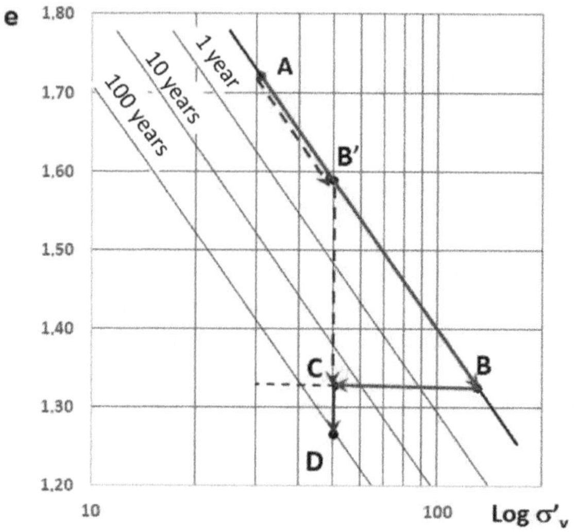

Figure 2.19. *Schematic diagram of the path taken by the effective vertical stress at a point in the residue below the axis of the embankment during the construction phase as per Bjerrum (1967). For a color version of this figure, see www.iste.co.uk/villard/ geosynthetics.zip*

Designing the geosynthetics for the deliberately partial punching phase: to obtain an "instant" vertical displacement of the embankment around 2 m, calculation of the force leading to the rupture of an embankment of infinite length yielded a height of 5 m. Although rather approximate, this methodology appeared adequate for the design of both the experimental structure and the geosynthetics.

In this first phase, the geosynthetics had to ensure the integrity of the embankment during the voluntary punching. Two types of potential failure were identified:

– localized failure of the foundation soil along a circular failure plane;

– failure of the geosynthetic during the overall settlement of the embankment as a result of the tensile stress generated by horizontal displacements of the foundation soil.

In practice, design vis-à-vis localized failure can easily be done using the usual geotechnology software through the method of slices. This makes it possible to integrate the ultimate tensile strength ($R_{t;d}$) of the geosynthetic in accordance with the Eurocodes. The calculation yielded a value $R_{t;d}$ = 120 kN/m for an embankment slope of 3:1. The reduction factors of the geosynthetics would be finalized at the end of the design phase to take the impact of the different phases into account.

The design vis-à-vis the rupture during the overall punching of the embankment is generally carried out using the finite element method, for which the choice of soil behavior parameters is crucial. In the present case, the lack of representative data and the heterogeneous nature of the foundation soil made this type of modeling very questionable. In addition, calculations carried out on structures on relatively homogeneous soils seemed to show that this type of failure is less critical than localized failure (Rowe 1984).

Design of geosynthetics for the consolidation and unloading phase: during the consolidation phase, the progressive increase in soil strength did not lead to an increased mobilization of the geosynthetics. This was the same for the unloading and construction phase of the pavement.

As the three phases of construction were carried out over a short period, less than a year, the use of short-term reduction factors for the reinforcement geosynthetics is justified. Table 2.8 presents the respective reduction factors for the three-phase solution with unloading (A, B, C, D) and the "classical", single-phase solution (A, B', D) (Figure 2.19).

The factors presented here correspond to the characteristics of the site:

– The in situ soil and that of the embankment were sandy materials that are not very aggressive, and the compaction intensity was low to average.

– The in situ soil and that of the embankment are materials that are, a priori, not very chemically aggressive (pH 8–9).

– The geosynthetic available at the time of construction was a woven polypropylene geotextile. It can be seen that the three-phase construction (and unloading) using reinforcement geosynthetics offers an advantage over the conventional solution by enabling:

- a considerable reduction in the rate of secondary consolidation of the soil, thereby preventing significant maintenance costs for the roadway;

- a significant reduction in the reduction factors of the reinforcement geosynthetics.

Reduction factors	Three-phase solution	Conventional solution
Γ_{CR}*	3 (1)	6 (1)
Γ_{CH}*	1.05 (2)	1.3
Γ_{ID}*	1.1	1.1
Γ_{CR}* . Γ_{CH}* . Γ_{ID}*	3.5	8.6

(1) These values correspond to the default values in the absence of any test data from an accredited laboratory.

(2) This value is safe as the duration of the actual mobilization of the geosynthetic, about one year at most, is much shorter than the 50 years taken into account for the value of 1.05.

Table 2.8. *Reduction factors to be used for a polypropylene geosynthetic in the absence of data from the manufacturer (see section 2.3.2.4)*

2.4.1.3. *Observations on the structure*

In 1986, when the structure was to be built, it was decided that an experimental, instrumented structure would first be built in the center of the basin. This was done by taking into account the absence of representative data and the heterogeneous nature of the foundation solution, as well as the tricky design of this kind of solution. The objective was to verify the feasibility of carrying out a controlled voluntary punching, as well as to confirm the design of the geosynthetics. The construction of the experimental embankment, 5 m high, was carried out in 15 days (three days for the last 3 m); the observed settlement was 1.8 m, with significant pore-pressures of between 80 and 100 kPa in the residue under the embankment.

During this initial phase, the embankment displaced vertically in a manner which was monolithic in nature; the deformations measured in the geosynthetics were of the order of 1.56 to 2.56%. It will be noted that these correspond to tensile stresses that are much smaller than those taken into account in the design (around half). During the consolidation and unloading phases, the relative variations in the deformations of the geosynthetics did not exceed 0.005%, which confirmed that there was virtually no additional mobilization of the geosynthetics.

Figure 2.20 depicts two aerial views of the structure. The photo on the left is from 1986, during the construction of the motorway, with the experimental embankment at the center. The photo on the right is from 2019, with the landfill dikes marked in red and the A26 highway in yellow. Figure 2.21 is a view from the side of the motorway in the north-west direction, that is, in the direction of the center of the former landfill.

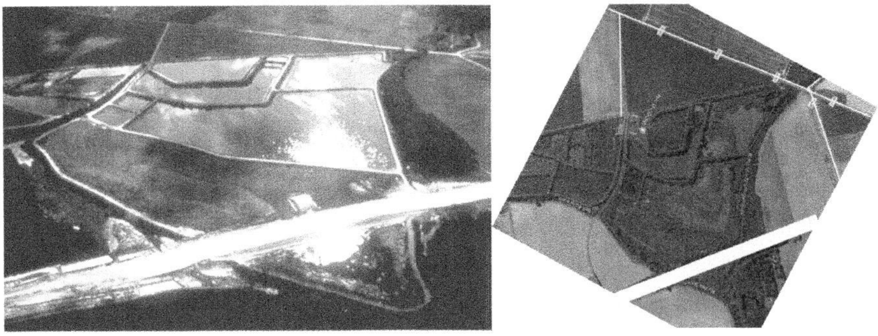

a) b)

Figure 2.20. *Aerial views of the structure in (a) 1986 and (b) 2019 (source: www.google.fr/maps). For a color version of this figure, see www.iste.co.uk/villard/geosynthetics.zip*

Figure 2.21. *View of the edge of the highway running north-west, that is, in the direction of the center of the former beet-washing site, in 2019 (source: www.google.fr/maps). For a color version of this figure, see www.iste.co.uk/villard/geosynthetics.zip*

2.4.2. *Reinforcement of an embankment on soft soil: examining circular failure (A64, the Guiche test embankment, France)*

Beyond the specific case of crossing over a very soft zone by using voluntary punching, the use of geosynthetic reinforcement at the base of an embankment has become a more widespread solution to the tricky problem of stabilizing embankments on compressible soils in the short term.

This technique mobilizes a resisting moment to increase the safety; in this time the soft soil consolidates, leading to an increase in cohesion sufficient to ensure long-term stability. Experience shows that the use of geosynthetics generally makes it possible to reduce the number of construction phases initially foreseen. This behavior has been demonstrated over the years in multiple test structures, for example, the experimental AIT structure in Bangkok (Long et al. 1997; Bergado et al. 1998), which was constructed in a very short time, that is without any appreciable consolidation of the underlying soil.

Table 2.9 shows the diagram of the undrained cohesions measured in the soft soil as a function of the depth.

Depth (m)	0	1	2	3	6	8	10	12	14
c_u (kPa)	14	12.3	10.7	9	9	11	17	21	25

Table 2.9. *Diagram of the undrained cohesions as a function of depth*

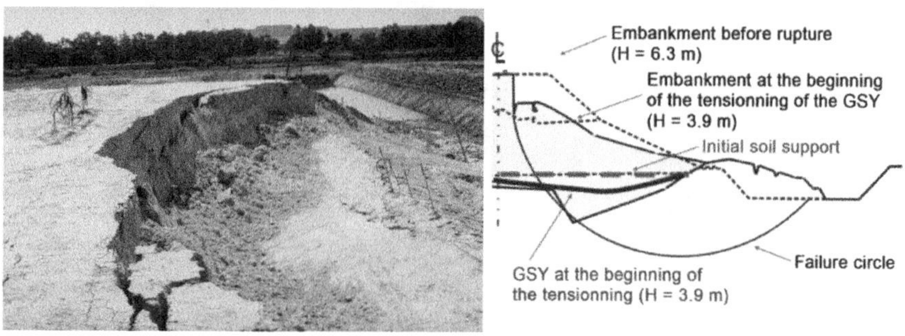

Figure 2.22. *Test embankment in Bangkok (Long et al. 1997). For a color version of this figure, see www.iste.co.uk/villard/geosynthetics.zip*

It was possible to show that the maximum height of the embankment without geotextiles was 3.9 m. However, thanks to reinforcement using geosynthetics with a tensile strength ($R_{t;k}$ = 200 kN/m), it was possible to raise this to a height of 6.3 m before failure occurred (Figure 2.22).

In order to study the approach used for planning and designing geosynthetic-reinforced embankments on soft soil, let us discuss a full-scale test structure that allowed engineers to verify that the presence of the reinforcement geosynthetics does not modify the most probable failure mechanism, and that the usual design, based on stability vis-à-vis circular failure, is much more critical. This was the case in the test structure at Guiche (Quaresma 1992; Delmas et al. 1994).

2.4.2.1. *Introducing the problem statement for the structures*

The test embankment at Guiche, located in the Adour valley, aimed to study the failure behavior of this kind of structure. The soil on the site consisted of a layer of mud, overlain by a layer of slightly over-consolidated clay. The thickness of the compressible layer was around 25 m and the geotechnical characteristics are presented in Table 2.10.

Depth (m)	w (%)	Ip	γ (kN/m^3)	e_0	c_u (kPa)	Cs	Cc	γ'_p (kPa)
0–1	31–41	37	18.0	1.1	43	–	–	–
1–3	56–80	56	16.6	1.6	31	0.03	0.6	46
3–7	90–247	–	13.5	4.0	30	0.15	1.5	55
7–18	46–60	34	17.0	1.3	32	0.044	0.7	94
18–24	30–50	–	17.5	1.1	–	–	–	–

Table 2.10. *Geotechnical characteristics of the soft soil on which the Guiche test embankment was constructed*

The embankment was constructed on the site using a lateral stabilizing berm with a height of 4 m, which made it possible to induce lateral failure of the embankment. The embankment was made of sand overlain by a clayey gravel up to a height of 8.75 m, at which the failure point was reached.

2.4.2.2. *Observations on the structure*

The measurements carried out using the comprehensive instrumentation on the embankment and on the reinforcement geosynthetic at the base of the embankment are presented in Figures 2.23 and 2.24.

The horizontal displacements measured at foundation level were large to the right of the embankment crest; they reached 19 cm for an embankment height of 8 m, while they remained small at the base of the slope. This agrees with the measurements carried out on the geosynthetic, as shown in Figure 2.23. It has been measured that all the horizontal displacements remain smaller than 1 cm up to an embankment height of 7 m and only exceed 5 cm in the final phase.

The failure obtained at H = 8.75 m is a rotational failure with the formation of a 2 m bulge at the toe and a step at the head of 2.5 m (Figure 2.25).

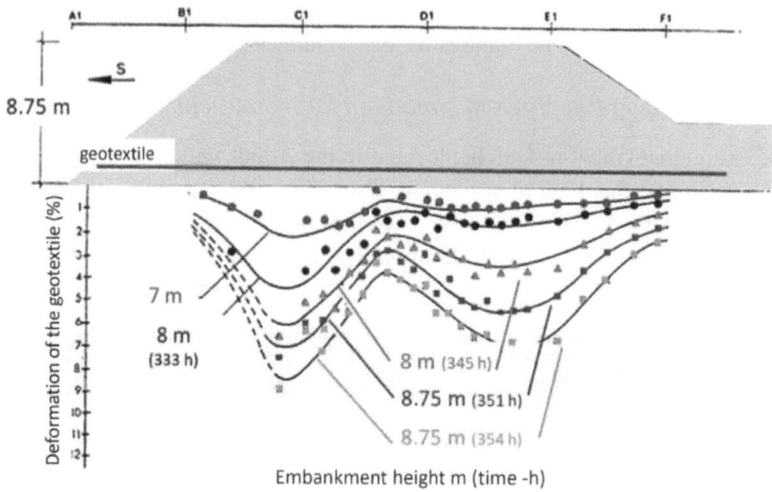

Figure 2.23. *Development of the distribution of deformations of the geotextile during construction (at the level of the geosynthetic placed on a sand bed). For a color version of this figure, see www.iste.co.uk/villard/geosynthetics.zip*

From the point of view of the observed behavioral mechanisms, the following can be observed:

– the delayed mobilization of the geotextile in tension when the embankment height was greater than 7 m; thus, the mobilization of the geotextile should be studied for in-service conditions with smaller heights;

– the superimposition of two mechanisms: the large soil-displacement above a height of 7 m and the rotational slip of the slope. This latter phenomenon predominates, but occurs without any perceptible starting point.

The deformation measurements of the geosynthetic to the right of the failure plane ($\varepsilon = 7.5\%$), for an embankment height of 8.75 m, correspond to a force of 185 kN/m. However, for the same embankment height, a greater deformation ($\varepsilon = 9.0\%$) was measured to the right of the top of the slope, corresponding to a force of 192 kN/m.

Considering that the measurements were not carried out continuously but intermittently over time, it would be reasonable to think that the actual force mobilized is at least equal to the highest measured force, that is $R_{mob} \geq 192$ kN/m (the tensile failure observed on site; see Figure 2.25). It should be noted that the value of the mobilized force is quite close to an estimate of the strength of the damaged geosynthetic: $R_{t;k}/\Gamma_{ID}^{*} = 213$ kN/m, with $R_{t;k} = 235$ kN/m and $\Gamma_{ID}^{*} = 1.1$.

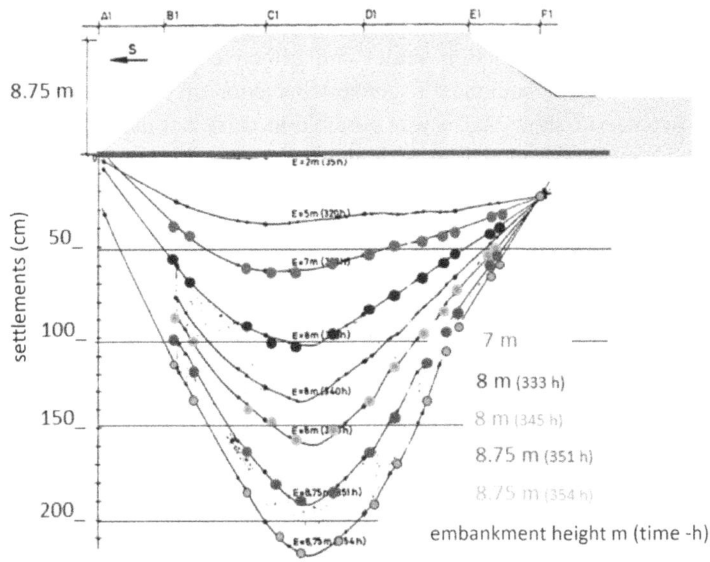

Figure 2.24. *Development of the vertical displacements under the embankment during construction n (at the base of the embankment). For a color version of this figure, see www.iste.co.uk/villard/geosynthetics.zip*

a) b)

Figure 2.25. *Rotational failure of the embankment: (a) view of the step at the head; (b) view of the failure of the geosynthetic after clearing the area. For a color version of this figure, see www.iste.co.uk/villard/geosynthetics.zip*

2.4.2.3. *Planning and design elements*

The finite-element calculation under the effective stresses enabled a good approximation of the embankment's general behavior up to a height of around 6.5 m. This seemed to show that it was possible to carry out the design for service states using finite-element modeling, as long as adequate site data on the parameters of the foundation soils are available.

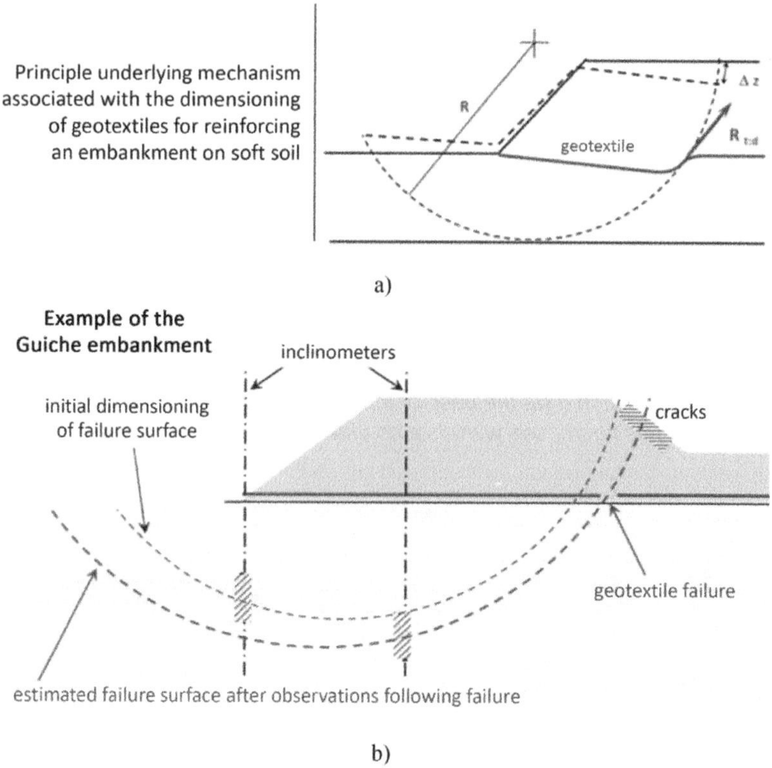

Figure 2.26. *(a) Principle underlying the mechanism associated with the design of geotextiles reinforcing an embankment on soft soil; (b) application to the Guiche embankment: comparison of the failure surface from the initial design with the surface derived from observations after failure. For a color version of this figure, see www.iste.co.uk/villard/geosynthetics.zip*

However, in present-day structures, the design is generally carried out using a limit-equilibrium approach (slip-circle failure) that makes it possible to incorporate a tensile force from the geotextile at the level of the estimated failure surface.

Figure 2.26 depicts the failure surface obtained in the preliminary calculations, and the surface derived from observations after failure has occurred.

The rupture calculations, for the surface derived from the observations, allows evaluation of the geosynthetic tensile strength $R_{t;d}$ for an embankment height of 8.75 m ($R_{t;d} \geq 240$ kN/m).

For operational design corresponding to this height (8.75m) for example, it would be appropriate to choose a geosynthetic with a characteristic tensile strength ($R_{t;k}$) calculated from equations [2.1] and [2.2] (section 2.3.2.3).

That is:

$$(R_{t;k}) = \gamma_{M;t} \,.\, \Gamma_{CR}^* \,.\, \Gamma_{CH}^* .\, \Gamma_{ID}^* .\, R_{t;d} = 330 \text{ kN/m}$$

where:

– $\gamma_{M;t} = 1.25$ (standards for geosynthetic-reinforcement);

– $\Gamma_{ID}^* = 1.1$ (if using a product similar to the experimental worksite and taking into account the damage tests carried out on this kind of product; the choice of a different product type would require this factor to be adjusted);

– Γ_{CR}^* and $\Gamma_{CH}^* = 1.0$ (for very short-term structures, with the geosynthetic being placed under stress for a period shorter than six months).

It can be seen that the value of ($R_{t;k}$) is much greater than the tensile strength value for the geosynthetic used for the experimental embankment at Guiche ($R_{t;k}$ = 235 kN/m), which was built to allow the observation of the rupture mechanism and which was deliberately under-dimensioned for this reason.

– Limit equilibrium method (displacement failure): with respect to the operational design of an embankment on soft soil, it is useful to verify, in parallel, the stability opposite the risk of displacement failure for the embankment. This can be done, for example, by using the conventional methods used for surface foundations with appropriate weighting.

Section 2.3.2. can provide more information.

– Method using deformations associated with consolidation settlements: observations from current structures on soft soils, reinforced with geosynthetics, show that the use of geosynthetics to ensure stability against circular failure does not result in adequate rigidification. It is therefore not possible to envisage a modification in settlements (in amplitude and distribution) compared to the situation without any reinforcement at the base of the embankment (taking into account the delayed mobilization of the reinforcement, Figure 2.23) (Rowe 1984; Delmas et al. 1987). The settlement of the reinforced embankments on soft soil is computed in the same way as for "conventional" embankments without reinforcement.

2.4.3. *Stabilization of a thin layer on a slope: the case of stabilization of a geosynthetic lining system on a slope*

The justification for the design elements for stabilizing a thin layer on a slope using geosynthetics is covered in the NF G38-067 (AFNOR 2017) standard. The working principle is shown in Figure 2.27. The stability of the structure depends on the internal shear strength of the covering layer and on the resistance to sliding of the different interfaces. The overlying structure must not lead to tensile forces in the sealing layer.

The overlying structure leads to destabilizing shear stressing (H_d) and to an interface shear strength ($R_{f;d}$) (Figure 2.28). If this shear strength is insufficient, it must be complemented by a resisting force from the reinforcement geosynthetic ($T_{max;d}$) anchored at the crest.

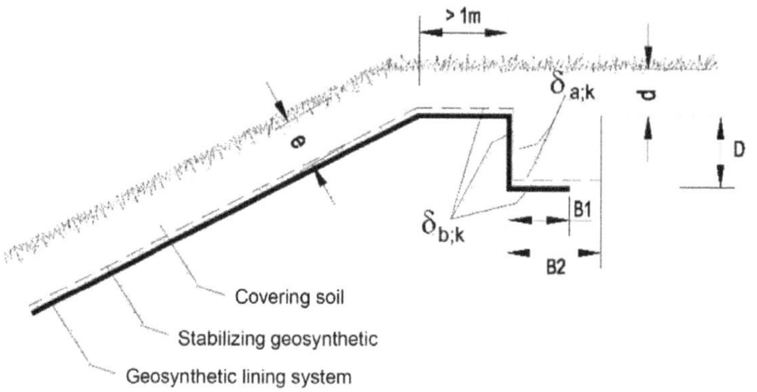

Figure 2.27. *Working principle of a thin layer on a slope (as per NF G38-067 (AFNOR 2017)) with $\delta_{a;k}$ as the characteristic value of the minimum angle of friction for the interfaces above the reinforcement geosynthetic, and $\delta_{b;k}$ as the characteristic value of the minimum angle of friction for the interfaces located below the reinforcement geosynthetic*

The NF G38-067 (AFNOR 2017) standard presents two well-documented design examples:

– A lining system (LS) with a drainage geocomposite on an HDPE sealing layer on a geotextile support; the slope is inclined at an angle of 27°, with a friction angle of 10° between the sealing layer and the geotextile support. This example provides a detailed study of the stability of the cover layer.

– A lining system with a geotextile on a PVC sealing layer on a geotextile support; the slope is inclined at an angle of 16° with a friction angle of 18° between the upper geotextile and the sealing layer. This example provides a detailed study of the stability of the cover layer and details the calculation of the tensile force ($T_{max;d}$) to be mobilized within the upper geotextile to prevent sliding of the soil layer, as well as the calculation of the characteristic tensile force ($R_{t;k}$) required for the upper geotextile. The example also shows calculations for the anchoring trench for the uptake of tensile forces in the upper geotextile.

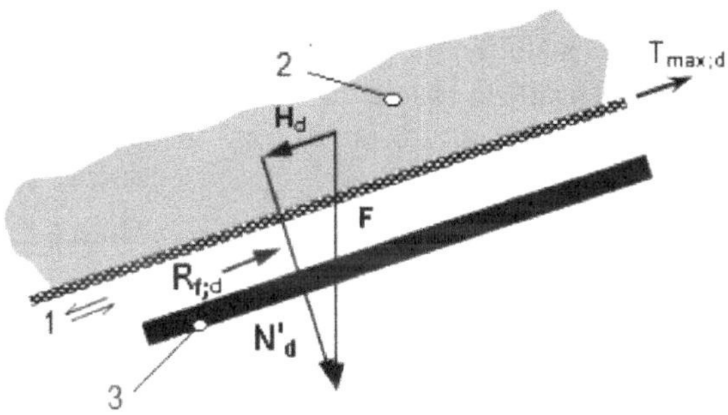

Figure 2.28. *Forces at the level of the cover layer (as per NF G38-067 (AFNOR 2017)) at the level of the slip plane under consideration (1), with the cover soil (2), on the geosynthetic lining system (LS) (3). For a color version of this figure, see www.iste.co.uk/villard/geosynthetics.zip*

OBSERVATIONS ON FIGURE 2.28.– *$T_{max;d}$ is the calculated value of the maximum tensile force in the reinforcement geosynthetic; H_d is the computed value for that component of the resultant of the actions applied to the body of fill material which is tangential to the slip plane; $R_{f;d}$ is the calculated value of the ultimate interaction strength at the level of the interface with lowest friction in the lining system.*

Both these examples follow the principles of the standard and are compliant with the Eurocodes. They are thus useful references. However, this chapter will now focus on the case of structures subjected to water pressure on either side of the sealing layer, which is not covered in the standard.

2.4.3.1. *Presenting the problem statement for the structure*

The example presented here is inspired by the study carried out by Soyez et al. (1990). This has been transposed so as to make it compliant with Eurocode design principles and by using the factors recommended by the NF G38-067 (AFNOR 2017) standard for structures without water. Figure 2.29 shows the geometry of the cross-section under consideration: the upstream height, H, is taken as 10 m, the water in the basin has a depth, h_i, of 5 m, the thickness of the protective layer is 1 m, and the slope is $\beta = 15°$. Table 2.11 gives the geotechnical characteristics of the soils and the different soil–geosynthetic and geosynthetic–geosynthetic interfaces.

It is interesting to see how an unexpected inflow of water upstream of the geomembrane affects the stability of the multilayer, and to evaluate the force mobilized in the upper geotextile to ensure the stability of the protective layer.

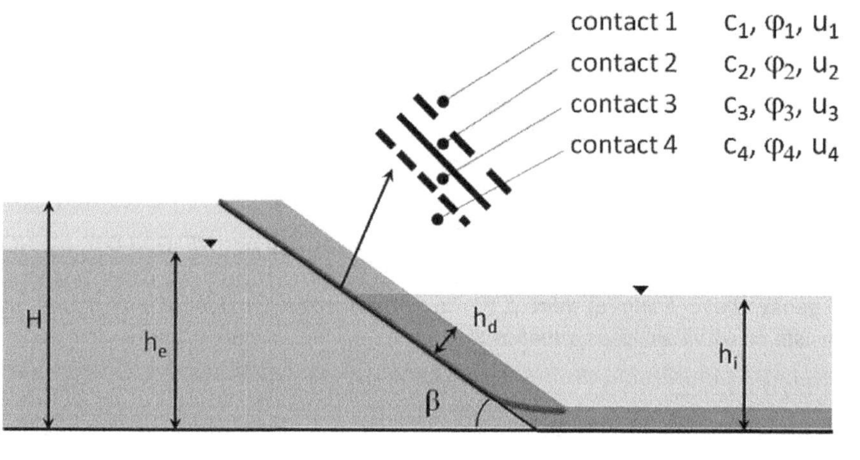

Figure 2.29. *Cross-section studied: the pore-pressures are governed by the hydrostatic equilibrium on either side of the geomembrane: u_1 and u_2, related to h_i within the basin; u_3 and u_4, related to h_e outside the basin. For a color version of this figure, see www.iste.co.uk/villard/geosynthetics.zip*

	γ (kN/m³)	c' (kPa)	φ' (°)
Upstream soil	19	0	35
Soil in the cover layer	19	0	35
Contact 1: cover soil – upper geotextile			29
Contact 2: upper geotextile – geomembrane			18
Contact 3: geomembrane – lower geotextile			18
Contact 4: lower geotextile – upstream soil			29

Table 2.11. *Geotechnical characteristics of the soils and of the different soil–geosynthetic and geosynthetic–geosynthetic interfaces*

2.4.3.2. *Proposed solution: concept, design elements and observations*

The computation is carried out using the software introduced by Soyez et al. (1990). This is a calculation of the geotechnical stability based on the method of slices, which allows integration of up to three geosynthetics lining systems and 10 interfaces in contact with the geosynthetics.

The weighting of the actions, soil parameters and loads is taken into account in the same way as in the procedure described in the standard.

Table 2.12 shows the development of the lining systems–upstream soil interface and the force to be mobilized in the upper geosynthetic ($R_{t;d}$) depending on the gradual rise in the upstream water level (h_e), as well as the number of the associated contact surface.

	h_e = 5.0 m	h_e = 7.5 m
"Lining system–upstream soil" interface	–	Lifting of the geomembrane
$R_{t;d}$ (contact no.)	77 kN/m (contacts 2 and 3)	– (contact 3)

Table 2.12. *Development of the lining system–upstream soil interface and the force to be mobilized in the upper geotextile ($R_{t;d}$) as a function of the gradual rise in the upstream water level (h_e)*

The calculations show that for an upstream water height of 7.5 m, the upstream interstitial pressure is greater than the total stress applied to the membrane. This results in the geomembrane being lifted for over 30% of the length of the slope. This is unacceptable and confirms the need for adequate upstream drainage.

Figure 2.30 presents the effective normal stress under the geomembrane (σ'_{ne}) (dashed line) compared to the total normal stress on the geomembrane. It shows that the resultant is positive for the full length of the slope. This assumes that the difference between the upstream water height and the height in the basin stays less than the effective normal stress applied by the overlying layer (lifted).

The characteristic tensile strength of the upper geotextile ($R_{t;k}$) is calculated in accordance with the standard NF G38-067 (AFNOR 2017), as presented in section 2.3.2.2.

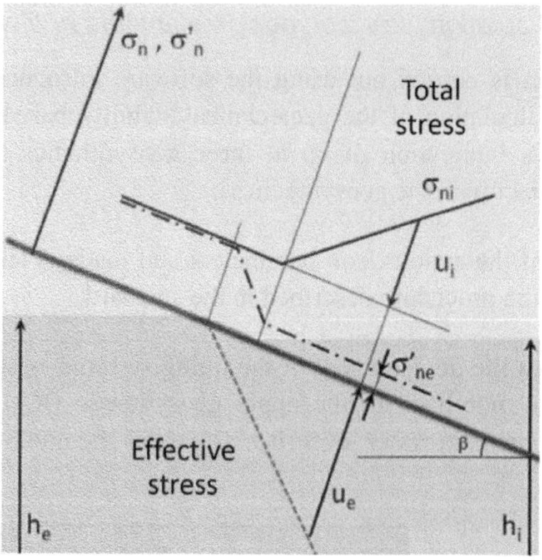

Figure 2.30. *Diagram showing the effective normal stress (σ'_{ne}) under the geomembrane compared to the total normal stress (σ_{ni}) on the geomembrane. For a color version of this figure, see www.iste.co.uk/villard/geosynthetics.zip*

It must also be ensured that the upstream drainage (and the maintenance of this drainage) makes it possible to guarantee that the upstream level remains smaller than the maximum predicted level (in the present case, $h_e \leq 5.0$ m).

Another problem that could affect the stability of the geosynthetic lining systems is that the cover layer may be more or less saturated in case of bad weather, or if there is no geosynthetic (e.g. drainage geocomposite including a geospacer) provided above the geomembrane, or if the geosynthetic is under-dimensioned.

This is a frequent cause for the sliding of the cover layer (Figure 2.31).

Let us consider the conditions at the level of the critical interface. Let δ_K be the characteristic value of the angle of friction at the interface.

Consider an infinite slope inclined at an angle β (Figure 2.32). Let z be the vertical thickness of the soil cover layer whose wet/saturated specific weight is equal to γ, and let z_w be the height of the water within the soil cover layer.

Figure 2.31. *Sliding of the cover layer on its support t. For a color version of this figure, see www.iste.co.uk/villard/geosynthetics.zip*

The interstitial pressure at the level of the critical interface is equal to:

$$u = \gamma_w.(z - z_w) = \gamma_w.(z - z_w).\cos^2\beta$$

and the corresponding hydraulic force, for an interface length L, is equal to:

$$U = u.L$$

The resultant, F, of the action of the cover layer on the interface can be split into:

– H_d tangential component: $H_d = \gamma.z.L.\sin\beta.\cos\beta$.

– N_d normal component: $N_d = N'_d + U = \gamma.z.L.\cos^2\beta$.

where N'_d is the effective component, normal to the soil-geomembrane interface:

$$N'_d = [\gamma.z.L.\cos^2\beta - \gamma_w.L.(z - z_w).\cos^2\beta]$$

Let $R_{f;d}$ be the ultimate resultant of interaction at the level of the critical interface:

$$R_{f;d} = N'_d.\tan\delta_k$$

The stability of the cover layer on the underlying geomembrane requires verification:

$$H_d < R_{f;d} \text{ with } R_{f;d} = N'_d \tan\delta_k = [\gamma.z.L.\cos^2\beta - \gamma_w.L.(z - z_w).\cos^2\beta].\tan\delta_k$$

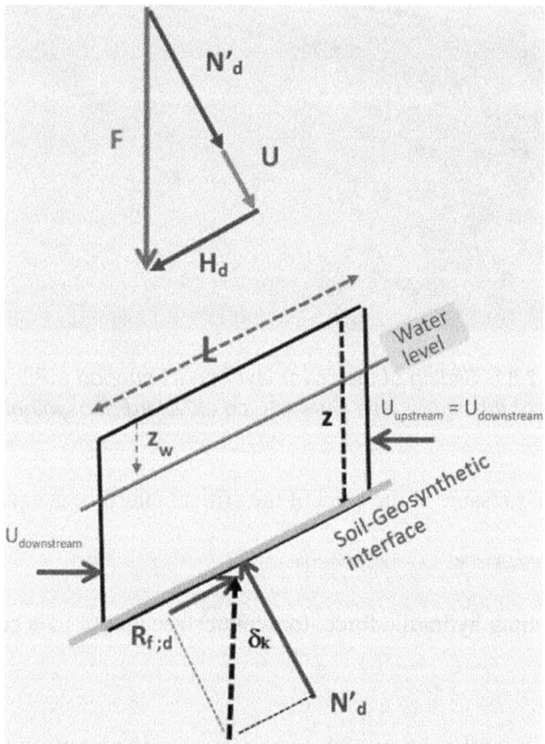

Figure 2.32. *Schematic diagram depicting the forces at the soil–geomembrane interface in the case of partial saturation of the cover layer. For a color version of this figure, see www.iste.co.uk/villard/geosynthetics.zip*

It can easily be seen that since H_d is fixed and the ultimate resultant of the interaction at the interface, $R_{f;d}$, decreases when the saturation of the soil layer increases (z_w decreases), any rise of the water level within the soil layer makes the stability increasingly critical.

The soil layer begins to slide when there is insufficient drainage during heavy rainfall.

On the other hand, if a drainage geocomposite above the geomembrane does not evacuate water from the soil cover layer, another solution is possible. This involves including an anchored reinforcement geosynthetic in the cover layer that is capable of resisting a tangential force T_d such that:

$$H_d < R_{f;d} + T_d$$

2.4.3.3. *Other examples of structures*

Examples of older structures still in operation and making use of this technique are presented in detail in Chapter 7:

– with upstream sealing geosynthetics on an earthwork dam (Ospédale 1979);

– with an upstream sealing complex on an earthwork dam, which caused problems during the construction as the friction between geosynthetics was overestimated (Aubrac 1986).

2.4.4. *Reinforcement of an embankment over potential subsidence area*

Given the risk of potential subsidence area with natural and/or anthropogenic cavities at a shallow depth, reinforcement geosynthetics are now relatively commonly used to prevent localized problems.

Many experimental and numerical studies have made it possible to understand the behavior of these geosynthetic-reinforced structures (Villard and Briançon 2008; Delmas et al. 2015; Al Heib et al. 2022).

Among other outcomes, these studies have allowed the publication of the French standard XP G38-065 (AFNOR 2020), relating to the planning and design of the geosynthetic reinforcement at the base of embankments on potential subsidence area, in the case of embankments on non-cohesive soils. The standard gives examples of design as per Eurocodes principles, in accordance with the serviceability limit states and ultimate limit states.

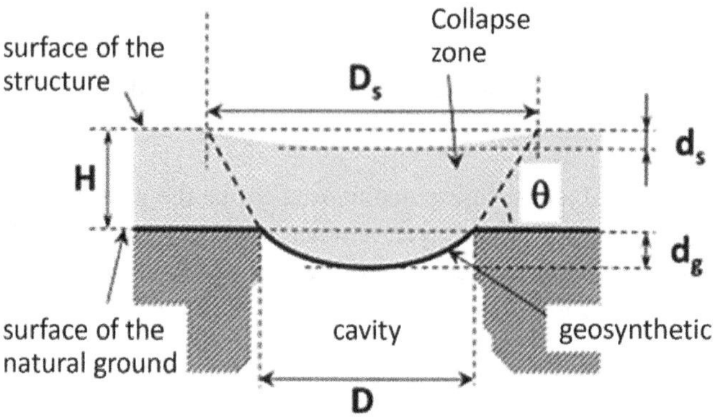

Figure 2.33. *Schematic cross-section of the structure with notations. In accordance with the standard, design will be carried out with $\theta = 90°$*

The rest of this section will show the influence of the different parameters that come into play. This will be done through a simple parametric study based on the example of a real structure, and it will be carried out in accordance with the standard. Figure 2.33 depicts a schematic cross-section of the structure with notations.

2.4.4.1. *Problem statement for the structure*

The example used is the structure presented by Riot et al. (2013), located on the A36 highway near Montbéliard (France). During the construction, several karst cavities were identified here, ranging in diameter from 0.70 to 0.80 m. The final structure can be considered as an embankment with a height of 1.00 m and a variable traffic load of 30 kPa.

It can be seen that the hypotheses (Table 2.13) regarding the soil behavior and the design method used in the following analysis are deliberately different from those used for the actual structure. This is done to simplify the discussion and comply with the XP G38-065 (AFNOR 2020) standard, published in 2020 after the structure was completed.

<table>
<tr><th colspan="6">Geotechnical characteristics</th></tr>
<tr><th colspan="2">Embankment</th><th colspan="2">Soil-geotextile interface</th><th colspan="2">Variable traffic load</th></tr>
<tr><td colspan="2">$\gamma = 21$ kN/m^3
$C_e = 1.03$
$\varphi_{sol} = 35°$
$c_{sol} = 0$
K = 1 (1)</td><td colspan="2">$\varphi_{soil/geo} = 26°$
$c_{soil/geo} = 0$
$u_0 = 1.5$ mm</td><td colspan="2">$Q_t = 30$ kPa</td></tr>
<tr><th colspan="6">Geometry and goals</th></tr>
<tr><th colspan="2">Embankment height</th><th colspan="2">For cavity size</th><th colspan="2">Goal for surface settlement</th></tr>
<tr><td colspan="2">Short term
H = 2.70 m
$H_{calculation} = 2.40$ m (1)
Long-term
H = 1.00 m</td><td colspan="2">D = 0.80 m</td><td colspan="2">$d_s = 2$ mm</td></tr>
<tr><th colspan="6">Characteristics of the geosynthetics</th></tr>
<tr><th colspan="3">Polymer</th><th colspan="3">Behavior over time (2)</th></tr>
<tr><td colspan="3">PET polyester ($\varepsilon_{max} = 10\%$)</td><td colspan="3">$\Gamma_{geo} = \Gamma_{CR}^* . \Gamma_{CH}^* . \Gamma_{ID}^*$
$C_{geo} = C_{CR}^* . C_{CH}^* . C_{ID}^*$
Short term $\Gamma_{geo} = C_{geo} = 1.77$
Long term $\Gamma_{geo} = C_{geo} = 2.3$</td></tr>
<tr><td colspan="3">(ELU) $\gamma_{M;t} = 1.25$</td><td colspan="3">–</td></tr>
</table>

(1) According to the standard XP G38-065 (AFNOR 2020), the calculation method used in the project in the standard is applicable for a ratio $1.5/4 < H/D < 3$. K value (Terzaghi method): K = 1.

(2) For simplicity's sake, the design is carried out directly from the reduction factors (Γ^*, the reduction factor for the strength of the geosynthetic and C*, the reduction factor for the force of the geosynthetic at the in-service strain); this is a reasonable approximation for the analysis which follows.

Table 2.13. *Basic design hypotheses*

2.4.4.2. *General design procedure*

The computations were carried out using the method developed by Villard and Briançon (2008), which makes it possible to carry out calculations for geosynthetic reinforcement over cavities in accordance with the standard by taking into account:

– circular or linear cavities;

– a non-cohesive embankment soil;

– a coefficient K = 1, or K_a based on the geometry for computing the normal stress on the geosynthetic above the cavity (Terzaghi method);

– an elastic behavior of the geosynthetic;

– a linear, elastoplastic soil/geosynthetic interaction;

– flat lateral anchoring;

– or trench anchoring (considering the geosynthetic to be clamped at the side of the trench).

The calculation is performed at the serviceability limit state for the reference diameter (here D = 0.8 m) for the desired surface settlement (here d_s = 2 mm). This gives the minimum stiffness, J, needed to ensure the desired surface settlement.

D = 0.8 m	Embankment height	SLS (d_s = 2 mm)		ULS (ε_g =10%)	
Parameter	(H) H_{calc}	J (kN/m)	T_{max} (kN/m)	J (kN/m)	T_{max} (kN/m)
Short term	(2.7) 2.4 m	66	9	48	11
Long term	1 m	1,994	21	54	16

Table 2.14. *Results for short-term and long-term conditions and a cavity diameter D = 0.8m*

The calculation is then carried out at the ultimate limit states for the reference diameter (0.8 m). This is a stability calculation for the rupture deformation of the GSY (here, ε_{max} = 10% for PET). In accordance with the XP G38-065 (AFNOR 2020) standard, partial factors are applied. This leads to the maximum tensile strength, T_{max}, required to ensure stability against failure. Table 2.14 summarizes the results for short-term and long-term conditions.

The design situation corresponds to the long-term serviceability limit state calculation with a stiffness J = 1,994 kN/m to satisfy the surface settlement requirement d_s = 2 mm; the corresponding deformation of the geosynthetic is ε_g = 0.46% and the settlement of the geosynthetic is d_g = 6.2 cm. Considering a polyester geosynthetic, with (ε_{max} = 10%), the product must meet the following specifications:

$R_{t;k} \geq 459$ kN/m $J_{\varepsilon = 0.46\%} \geq 1{,}994$ kN/m Type: polyester

For the long-term: $\Gamma_{geo} = C_{geo} = 2.3$

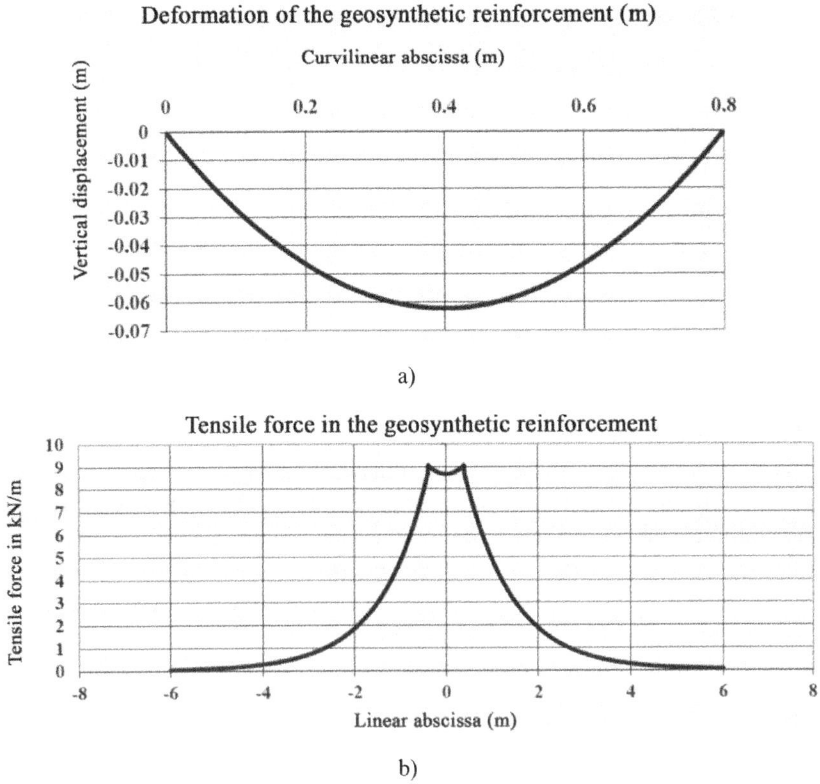

Figure 2.34. *Deformation of the geosynthetic above the cavity, and tensile force over the cavity and in the lateral anchoring. For a color version of this figure, see www.iste.co.uk/villard/geosynthetics.zip*

It can be seen that the rate of mobilization in tension of the product remains below 5% of the tensile strength of the geosynthetic.

Figure 2.34 depicts the deformation of the geosynthetic above the cavity, as well as the tensile force above the cavity and in the lateral anchoring. It can be seen that the horizontal anchoring forces are mobilized over a length of 6 m on either side of the cavity. Depending on the site constraints, anchoring can also be carried out through anchoring trenches (section 2.3.4.5).

2.4.4.3. *Analysis of the influence of variation in cavity diameter on the design*

Considering that the structure was built as per these geosynthetic specifications, it is interesting to attempt an assessment of the influence of a variation in the cavity diameter on the design. This variation may result from an underestimation of the reference diameter during the preliminary study, or from a change in the geological conditions over time. Similarly, it would also be interesting to look at how a gradual widening of the cavity may influence the behavior of the structure. The analysis was therefore carried out for diameters evaluated with respect to the reference diameter $D_{ref} = 0.80$ m:

– underestimation of the diameter: D_{ref} + 10%, D_{ref} + 20%, D_{ref} + 50%;

– gradual widening: 50% of D_{ref}.

Table 2.15 presents the results of the calculations for long-term conditions and a polyester geosynthetic ($R_{t;k} \geq 459$ kN/m, $J_{\varepsilon = 0.46\%} \geq 1{,}994$ kN/m, and a long-term $\Gamma_{geo} = C_{geo} = 2.3$).

It is evident that underestimating the diameter of the cavity results in surface settlements that are greater than those originally prescribed. However, these still remain within a range which (considering their amplitude) allows the project to go ahead, though this may suffer if the settlements necessitate a change in the speed of traffic. Instrumentation, along with appropriate auscultation and tracking, can permit measures to be anticipated that need to be implemented if a cavity develops whose diameter is larger than the reference diameter.

$R_{t;k} \geq 459$ kN/m, $J_{\varepsilon = 0.46\%}$ $\geq 1\,994$ kN/m	D_{ref}	D_{ref} + 10%	D_{ref} + 20%	D_{ref} + 50%	50% D_{ref}
Diameter of the cavity (m)	0.80 m	0.88 m	0.96 m	1.20 m	0.40 m
Surface settlement	0.2 cm	1.15 cm	2.1 cm	5.4 cm	0 cm
Max deformation of the GSY (ε_g)	0.46%	0.54%	0.63%	0.93%	0.14%
Percentage of mobilized tensile strength	4.6%	5.4%	6.3%	9.3%	1.4%

Table 2.15. *Results of variations in the diameter in long-term and short-term conditions for a polyester geosynthetic ($R_{t;k} \geq 459$ kN/m, $J_e = 0.46\% \geq 1{,}994$ kN/m, and in the long term $\Gamma_{geo} = C_{geo} = 2.3$)*

The study of the gradual widening of a cavity shows that for a cavity whose diameter is half that of the reference diameter, the geosynthetic deformation is very small (here 0.14%). It can therefore be seen that in order to detect the appearance of a cavity, it is often necessary to use specifically developed tools that can measure very small levels of deformation. Integrated systems consisting of detection, auscultation and warning modules have been developed recently, and these are associated with a specifically designed geosynthetic to allow the detection and measurement of very small deformations caused by the development of underground cavities.

2.4.4.4. *Another example of an older structure*

Another example of an older structure that is still operational and used this technique is presented in detail in Chapter 7. This is the "Trois Lucs à La Valentine" roadway (1990), in which geosynthetic reinforcement was used over cavities with lateral trench anchoring. When the system continued to function well even after the cavity rose to the level of the road, it was successfully renewed.

2.4.5. *Reinforced multilayer embankment*

The justification of the dimensioning of geosynthetic-reinforced embankments is given in NF G38-064 (AFNOR 2016) for reinforced inclined structures and slopes with a batter smaller than 1H/4V and NF P94-270 (AFNOR 2020) for vertical walls and walls with batter greater than 1H/4V. The procedures differ mainly in their calculation of internal stability. In the NF G38-064 (AFNOR 2016) standard, this is integrated with the so-called "mixed" stability.

These standards, which were published over a decade ago, are now in common use, just as geosynthetic-reinforced structures have also become common. That is why this chapter will now review the general design principles for this kind of structure, taking the example of the pre-existing structure (which is also discussed in Chapter 7). A special emphasis will be placed on how the evolution of standards can influence the design of structures.

2.4.5.1. *Review of the design procedure for multilayer structures reinforced with geosynthetics*

The procedure is similar for both standards and includes:

– the data collection required for the verification of the design:

- geometric and geotechnical data relating to the site and to the proposed materials for the embankment (permanent and/or temporary surcharges), climatic conditions, and hydrological and hydrogeological data,

- classification of the structure: defining the class of consequences and the geotechnical category as per the BS EN 1977-1 (BSI 2013a) standard,

- defining the lifespan of the structure;

– general principle for verification of the structure:

- selecting potentially unfavorable situations for the projects and the limit states to consider, both for the structure under construction and the structure in-service,

- study of the ultimate limit states of a representative profile:

a) verification of the geometry of the structure including verifying the external stability (sliding on the underlying soil and lack of bearing capacity of the underlying soil) and overall stability;

b) verification of the geosynthetics and their distribution: internal stability (NF EN 1977-1 (AFNOR 2013a)) and "mixed" stability. For each of these verifications, the appropriate partial factors will be applied to the actions, soil parameters, resistances and the model, in accordance with the standard;

- studying the serviceability limit states of a representative profile, which includes the verification of how the structure or structural elements function in normal service with regard to its ease of use by persons and its appearance:

a) consolidation settlements;

b) deformations originating internally in the construction and/or operation phase. For the latter, the deformability of the geosynthetics must be taken into account, especially the creep deformation from the end of the construction phase for the duration of the in-service life of the structure;

- in the French standards, this requirement is addressed by a coefficient that is integrated in the reduction factors of creep (Γ_{CR}* and corresponds to limiting the post-construction creep deformation to 3% for a vertical wall),

- in different countries in Europe, this post-construction creep deformation is limited to 1% for common vertical structures and 0.5% for load-bearing abutments.

Where geosynthetics are concerned, whether it is for the ULS or SLS, the general principles for evaluating behavior that were presented earlier in section 2.4 are applied. This includes using reduction factors to define product characteristics acceptable for the long term (see section 2.3.2.2).

2.4.5.2. *Presenting the problem statement for the structure*

The MS13 structure on the A20 highway between Foix and Tarascon (France) is a high, reinforced structure (21 m). This structure is discussed in detail in Chapter 7. It comprises two intermediary, 3 m berms. The typical cross-section of this structure is depicted in Figure 2.35. Given the length of the sheets, this must be considered as a single structure with a batter <1H/4V. The NF G38-064 (AFNOR 2016) standard was therefore used for the design of this structure. Further, for the operational design, it will be useful to verify the design of each block, in accordance with NF P94-270 (AFNOR 2020). The following analysis only considers the overall structure.

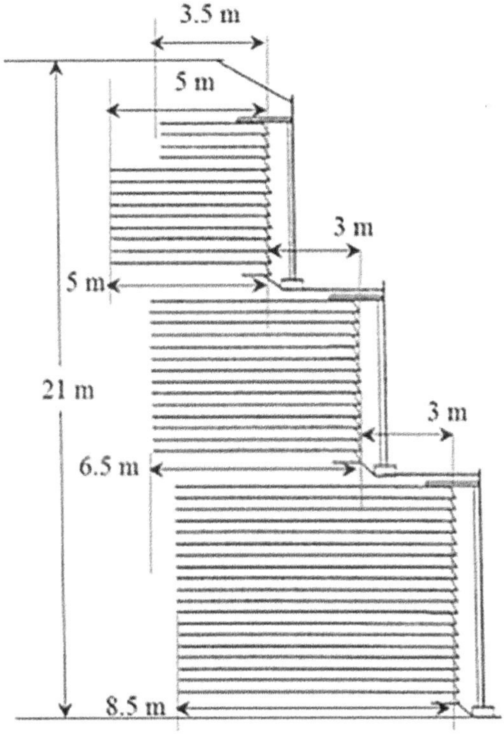

Figure 2.35. *Typical cross-section of the MS13 structure on the A20 between Foix and Tarascon (France)*

2.4.5.3. *Original design procedure*

In 1993, the design was carried out as per the recommendations and norms in effect, and by using the “displacement method” and respecting the French standards of the 1990s (Hérault et al. 1995). The main characteristics of the structure are given in Table 2.16.

2.4.5.4. *Design procedure as per the NF G38-064 standard: influence of evolving standards*

It is also interesting to note that from the time that the original design was made, the rules have evolved significantly, with the standards accounting for Eurocodes (NF G38-064 (AFNOR 2016) and NF P94-270 (AFNOR 2020)):

– In the 1990s, there were no specific factors, but only a global safety factor SF = 1.5 for the shear strength of the soil (φ', c').

– Using the NF G38-064 (AFNOR 2016) standard, the design is carried out at the ultimate limit states, using the partial coefficients for actions, soil parameters, resistance and the computational model.

Table 2.17 provides a concise comparison of the approaches.

Foix-Tarascon Highway, MS13			
Geometry			
MS13 structure			
Height	21 m		
Length	150 m		
Batter	74°		
	Block 1 (base)	**Block 2**	**Block 3 (top)**
Height	7 m	7 m	7 m
Batter	90°	90°	90°
Geosynthetics			
Type	**BidimRock (1)**		
Tensile strength (base/top)	400/300 kN/m	200/150/100 kN/m	75/50 kN/mm
Spacing	0.4 m	0.4 m	0.4 m
Length	8.5 m	6.5 m	5.0/3.0
Embankment soil			
c', φ'	0 kPa, 35°		

(1) Product made by Bidim Geosynthetics during this period.

Table 2.16. *The Foix-Tarascon highway – MS13: characteristics of the 21 m high, reinforced structure*

	1993	NF G 38-064 (AFNOR 2016)	
Unfavorable (favorable) permanent actions	1.0	γ_{GSup} (γ_{GInf})	1.35 (1.0)
Unfavorable (favorable) variable actions	1.0	γ_{QSup} (γ_{QInf})	1.5 (0.0)
Friction angle (φ')	1.5	$\gamma_{\varphi'}$	1.25
Cohesion (c')	1.5	$\gamma_{c'}$	1.25
Specific weight	1.0	γ_{γ}	1.0
Geosynthetic tensile strength	(1)	$\gamma_{M;t}$	1.25
Interface friction	1.5	$\gamma_{R;f}$	1.35
Coefficient for the computation method	(2)	$\gamma_{R;d}$	1.1
Maximum elongation of the geosynthetic	(3)	ε_{max}	≤ 3%

(1) Integrated into the reduction coefficient of the geosynthetic.

(2) Integrated into the overall safety coefficient SF = 1.5.

(3) Integrated into the calculation using the "displacement method".

Table 2.17. *Comparison of the original design approach and the approach in accordance with NF G38-064 (AFNOR 2016)*

Similarly, the approach using the allowable tensile strength of the geosynthetic changed with advances in our knowledge of polymers and geosynthetics tested for longer periods (creep, chemical degradation). Table 2.18 shows a comparison of the reduction factors from both approaches, showing the maximum reduction factors prescribed by the standards (1) and the usual values supplied for the principal polyester products in the market (2).

		1993		NF G 38-064 (AFNOR 2016)	
Reduction factors		**(1)**	**(2)**	**(1)**	**(2)**
Creep	Γ_{CR}*	2.3	2.3	3.0	1.92
Ageing	Γ_{CH}*	1.1	1.1	1.2	1.2
Damage during installation	Γ_{ID}*	1.5	1.1	2.15	1.22
Partial factor (review of Table 2.17)					
Geosynthetic tensile strength	$\gamma_{M;t}$	1.2		1.25	

(1) Prescribed maximum reduction factors (standards).

(2) Usual values supplied for the principal polyester products.

Table 2.18. *Comparison of factors applied to the geosynthetics in the original design approach and using approaches in accordance with the NF G 38-064 (AFNOR 2016) standard*

The calculations were carried out using the same hypotheses on the embankment soil ($\gamma = 20$ kN/m^3, $\varphi' = 35°$, $c' = 0$ kPa) and the traffic ($Q = 10$ kPa), and with the same software, using the "displacement method". It can be seen that the most critical circle in accordance with NF G38-064 (AFNOR 2016) is the same as in the design carried out in 1993, and passes through the toe of the wall, intercepting the geotextiles in the lower block.

Table 2.19 shows the comparison of the values of ultimate tensile strength.

		1993		NF G 38-064 (AFNOR 2016)	
		(1)	(2)	(1)	(2)
Ultimate tensile strength	$R_{t;d}$ (kN/m)	569	417	512	348

(1) Prescribed maximum reduction factors (standards).

(2) Usual values supplied for the main polyester products.

Table 2.19. *Comparison of the ultimate tensile strengths when following the approach in use in 1993 and for the approach in accordance with NF G38-064 (AFNOR 2016)*

The ultimate tensile strength of the geosynthetic chosen for the base of the wall, 400 kN/m, was close to the value of 417 kN/m, corresponding to the design calculation of 1993, while the requirement as per NF G38-064 (AFNOR 2016) would have only been 348 kN/m.

2.4.5.5. *Importance of installation on the appropriate behavior of geosynthetic-reinforced structures*

Unlike metallic reinforcement, when designing structures with geosynthetic reinforcement, the extensibility of the material must be taken into account to ensure that the structure behaves appropriately. This is accounted for in the standards, as shown in section 2.4.5.1, by the criterion limiting post-construction deformation due to creep. However, the standards do not give any specific recommendations on methods to implement in order to satisfy this performance requirement.

Chapter 7 discusses several examples of older structures that are still operational, constructed using different techniques. By observing these structures over the long term, it has been possible to derive recommendations for the successful construction of geosynthetic-reinforced structures, depending on the objectives of the structure:

– Prapoutel (1982): slope reinforced with wrap-round facing with increased interlayer thickness.

– Brides-les-Bains (1990): walls with cellular facing reinforced by geosynthetics.

– Les Hospices de France (1987): retaining structure constructed using a specific technique that allowed heavy compaction of the backfill and pre-stressing of the geosynthetic while also creating an aesthetically pleasing facing.

– Foix-Tarascon highway (1993): the MS13 retaining structure on the A20 between Foix and Tarascon.

– Maraval (1976): dam with a vertical downstream facing reinforced by geosynthetics with a concrete spillway using the reinforced structure as footing.

– Lixing (1984): using a geosynthetic-reinforced structure to stabilize a landslide.

It must be noted that at the time of publication of this book, these structures, which were designed according to the rules that were state-of-the-art at the time of their construction, have exhibited no significant deformations over time that could compromise their current use.

2.5. References

AFNOR (1992). NF P94-210, Renforcement des sols – Généralités et terminologie. Standard, AFNOR, Paris.

AFNOR (2016). NF G38-064, Utilisation des géotextiles et produits apparentés – Murs inclinés et talus raidis en sols renforcés par nappes géosynthétiques – Justification du dimensionnement et éléments de conception. Standard, AFNOR, Paris.

AFNOR (2017). NF G38-067, Géosynthétiques, géotextiles et produits apparentés – Stabilisation d'une couche mince sur pente – Justification du dimensionnement et éléments de conception. Standard, AFNOR, Paris.

AFNOR (2020a). NF P94-270, Calcul géotechnique – Ouvrages de soutènement – Remblais renforcés et massifs en sol cloué. Standard, AFNOR, Paris.

AFNOR (2020b). XP G38-065, Géosynthétiques, géotextiles et produits apparentés – Renforcement de la base des remblais sur zones à risques d'effondrement – Justification du dimensionnement et éléments de conception. Standard, AFNOR, Paris.

Al Heib, M., Delmas, P., Riot, M., Emeriault, F., Villard, P. (2022). Renforcement par géosynthétique pour la réduction des risques associés à un effondrement localisé. Report, Ineris, Paris.

Baraize, E., Chavaz, H.P., Delmas, P., Gendrin, P., Haza, E., Nancey, A., Gourc, J.P. (1995). Chargement d'un massif renforcé par géotextiles essais en vraie grandeur. CFG.

Bergado, D.T., Long, P.V., Delmas, P., Loke, K.H., Miura, N. (1998). Performance of geotextile reinforced embankment on soft Bangkok Clay. In *Sixth International Conference on Geotextiles, Geomembranes and Related Products*, Atlanta, 2, 747–754.

Bjerrum, L. (1967). Engineering geology of normally consolidated marine clays as related to the settlement of buildings. *Geotechnique*, 17(2), 883–113.

Bourdeau, Y., Kastner, R., Bollo-Kamara, N., Bahloul, F. (1990). Comportement en ancrage d'un géosynthétique enfoui dans un matériau bidimensionnel. In *5ème colloque Franco-Polonais de Mécanique des sols Appliquée*, Cambefort.

Bourdeau, Y., Ogunru, T., Lareal, P., Riondy, G. (1994). Use of strain gages to predict soil-geotextile interaction in pullout tests. In *5th International Conference on Geotextiles, Geomembranes and Related Products*, Singapore.

Briançon, L. and Villard, P. (2008). Design of geosynthetic-reinforced platforms spanning localized sinkholes. *Geotextiles and Geomembranes*, 26(5), 416–428.

BSI (1999). BS EN ISO 13431, Geotextiles and geotextile-related products – Determination of tensile creep and creep rupture behaviour. Standard, BSI, London.

BSI (2006). BS EN 14475, Execution of special geotechnical works – Reinforced fill. Standard, BSI, London.

BSI (2008). BS EN 15381, Geotextiles and geotextile-related products – Characteristics required for use in pavements and asphalt overlays. Standard, BSI, London.

BSI (2009). BS EN ISO 10321, Geosynthetics – Tensile test for joints/ seams by wide-width strip method. Standard, BSI, London.

BSI (2010). BS 8006-1, Code of practice for strengthened/reinforced soils and other fills. Standard, BSI, London.

BSI (2013a). BS EN 1977-1, Eurocode 7: Geotechnical design – Part 1: General rules. Standard, BSI, London.

BSI (2013b). BS EN 1977-2, Eurocode 7: Geotechnical design – Part 2: Ground investigation and testing. Standard, BSI, London.

BSI (2015). BS EN ISO 10319, Geosynthetics – Wide-width tensile test. Standard, BSI, London.

BSI (2016a). BS EN 13249, Geotextiles and geotextile-related products – Characteristics required for use in the construction of roads and other trafficked areas (excluding railways and asphalt inclusion). Standard, BSI, London.

BSI (2016b). BS EN 13250, Geotextiles and geotextile-related products – Characteristics required for use in the construction of railways. Standard, BSI, London.

BSI (2016c). BS EN 13251, Geotextiles and geotextile-related products – Characteristics required for use in earthworks, foundations and retaining structures. Standard, BSI, London.

BSI (2016d). BS EN 13253, Geotextiles and geotextile related products – Characteristics required for use in erosion control works (coastal protection, bank revetments). Standard, BSI, London.

BSI (2016e). BS EN 13254, Geotextiles and geotextile related products – Characteristics required for the use in the construction of reservoirs and dams. Standard, BSI, London.

BSI (2016f). BS EN 13255, Geotextiles and geotextile related products – Characteristics required for use in the construction of canals. Standard, BSI, London.

BSI (2016g). NF EN 13257, Geotextiles and geotextile related products –Characteristics required for use in solid waste disposals. Standard, BSI, London.

BSI (2016h) BS EN 13265, Geotextiles and geotextile-related products – Characteristics required for use in liquid waste containment projects. Standard, BSI, London.

BSI (2019). BS EN ISO 10722, Geosynthetics – Index test procedure for the evaluation of mechanical damage under repeated loading – Damage caused by granular material (laboratory test method). Standard, BSI, London.

BSI (2023). BS EN 17738, Geotextiles and geotextile-related products – Damage during installation procedure – Full scale test. Standard, BSI, London.

Delmas, P., Berche, J.C., Gourc, J.P. (1986a). Le dimensionnement des ouvrages renforcés par géotextiles : programme Cartage. *Bulletin de liaison des LPC*, 142, 33–44.

Delmas, P., Soyez, B., Lenglet, J., Puech, J.P., Buffard, J.Y. (1986b). Poinçonnement expérimental d'un remblai renforcé par géotextile sur une décharge industrielle. In *II International Conference on Geotextiles*, Vienna.

Delmas, P., Magnan, J.P., Soyez, B. (1987). New techniques for building embankments on soft soil. *Bulletin of the Public Works Research Center*, 323–356.

Delmas, P., Magnan, J.P., Quaresma, M., Gourc, J.P., Haïdar, S. (1994). Remblai expérimental sur sol mou renforcé par géotextile. In *XIII ICSMFE*, New Dehli.

Delmas, P., Villard, P., Huckert, A. (2015). Dimensionnement à court terme et à long terme de structure renforcée par géosynthétique sur cavités potentielles : prise en compte de la sécurité. In *10ème Rencontres Géosynthétiques*. CFG, La Rochelle.

Elias, V., Salman, A., Goulias, D. (1988). The effect of pH, resin properties, and manufacturing process on laboratory degradation of polyester geosynthetics. *Geosynthetics International*, 5(5), 460–490.

Gorniak, J., Villard, P., Delmas, P. (2013). Analytical design method to determine the vertical loading capacity of geosynthetic tubes filled with granular material. *International Symposium on Design and Practice of Geosynthetic-Reinforced Soil Structures*, 5, 406–415.

Gourc, J.P., Villard, P., Matichard, Y. (1992). Pull-out behaviour of reinforcements. Centrifuge tests and theoretical validations. In *International Symposium on Earth Reinforcement Practice*, Kyushu.

Hérault, A., Mieussens, J.C., Nancey, A., Dalbion, J., Delmas, P. (1995). Les géotextiles de renforcement dans les ouvrages de soutènement : une technique éprouvée – MS13 RN20 Foix Tarrascon sur Ariège. CFG.

ISO (2020). ISO TS 13434, Geosynthetics – Guidelines for the assessment of durability. Standard, ISO.

Lawson, C. (2006). Geotextile containment for hydraulic and environmental engineering. In *8th ICG*. Giroud Lecture, Yokohama.

Long, P.V., Bergado, D.T., Balasubramaniam, A.S., Delmas, P. (1997). Localized mobilization of geotextile reinforcement force at failure surface. In *Fourteenth International Conference on Soil Mechanics and foundation Engineering*, Hamburg.

Müller-Rochholz, J. (2003). Modellvarianten zur Simulation von Einbaubeschä-dingungen. *tBu Newsletter*, 11(2).

Nait-Ali, L. and Freitag, N. (2014). Hydrolysis assessment methodology for polyester yarns used in soil reinforcements. Part 1: Chemical model. In *X*ème *Cong. Int. Geosynthetics*, Berlin.

Nait-Ali, L., Thomas, R., Freitag, N. (2014). Polyvinyl alcohol physical behaviour and chemical durability. In *X*ème *Cong. Int. Geosynthetics*, Berlin.

Quaresma, M. (1992). Comportement et modélisation d'un remblai sur sols mous renforcé par géotextile. PhD Thesis, Université Joseph Fourier, Grenoble.

Riot, M., Rouffiac, M., Auray, G. (2013). Chantier de l'autoroute A36 : étude de cas de renforcement par géosynthétique sur zones à risques de cavités karstiques. In *9*èmes *Rencontres Géosynthétiques*.

Rowe, R.K. (1984). Reinforced embankments: Analysis and design. *ASCE Journal of Geotechnical Engineering*, 110(GT2), 231–246.

Soyez, B., Delmas, P., Herr, C., Berche, J.C. (1990). Computer evaluation of the stability composite liners. In *4th Int. Conf. on Geotextiles, Geomembranes and Related Products*, La Haye.

Villard, P. and Briançon, L. (2008). Design of geosynthetic reinforcements of platforms subjected to localized sinkholes. *Can. Geotech. J.*, 45(2), 196–209.

Villard, P. and Chareyre, B. (2004). Design methods for geosynthetic anchor trenches on the basis of true scale experiments and discrete element modelling. *Can. Geotech. J.*, 41(6), 1193–1205.

3

Numerical Models Specific to Reinforcement

Audrey HUCKERT[1] and Francis TANO[2]

[1] *Laboratoire 3SR, Université Grenoble Alpes, France*

[2] *MERAMO, Chatenay-Malabry, France*

It is not self-evident that numerical modeling will be used for the design of geosynthetic reinforcement, which uses analytical methods. However, in the case of complex geotechnical structures where the interfaces between the reinforcement and soil may multiply, a numerical approach makes it possible to estimate the stresses induced in the reinforcement and to undertake its design.

3.1. Introduction

3.1.1. *The place of numerical modeling in designing geosynthetics*

The numerical modeling of a reinforcement geosynthetic is a complex problem due to the definition of the type of elements that must be used to accurately describe the mechanical behavior of the reinforcement and its interface with the soil. Geotextiles are thin, have low bending stiffness, and are made of flexible fibers, while geogrids are constituted by more or less stiff ribs distributed more often on rectangular or triangular meshes. Depending on the type of products, different stiffness of fibers or ribs can be used in different directions so that the behavior of such a product is highly anisotropic.

Geosynthetics from Yesterday to Today,
coordinated by Pascal VILLARD.

Current standards and recommendations on the design of geosynthetic reinforcements are rather imprecise with regard to numerical modeling. Being more precise is most often left to the persons carrying out the modeling who have only general recommendations and their own expertise to rely on. To accurately describe the specific behaviour of geosynthetic reinforcements in the numerical model, the following must be taken into account:

– a description of its tensile behavior (generally indicated on the manufacturer's data sheet) obtained from standardized characterization tests;

– stiffnesses under load compatible with the structure to be reinforced. In the case of woven or nonwoven geosynthetics, it is possible to define several, specific directions of reinforcement;

– low or no compressive strength, so that only tensile forces develop in the plane of the geosynthetic;

– a plane, frictional anchoring surface between the geosynthetic and the soil;

– a definition of the soil–geosynthetic interface based on experiments to characterize the material (inclined plane, specific shear box, pull-out test, etc.). This interface is subject to a set of complex mechanisms that are not completely understood, such as friction, the rolling of soil particles, or the interlocking of soil particles in the geogrid mesh or geosynthetic fibers;

– a modeling method that allows the geosynthetic to be deformed as a membrane, for example, through the use of appropriate and sufficient structural elements.

It is also the fact that for any geotechnical structure, numerical modeling makes it possible to complement analytical design by estimating the displacements within a complex structure.

3.1.2. *General information: continuous modeling*

Outside of the research context, finite modeling methods are the most widely used due to the ease of adapting existing software and the relatively high computing speed.

3.1.2.1. *General principles underlying continuous modeling*

Continuous modeling is a tool that can predict the behavior of one or more physical phenomena in a medium, thanks to the numerical resolution of the mathematical equations that describe these behaviors. The medium here is taken to be "continuous", that is, a domain within which the properties of the medium are continuous and, therefore, described by continuous mathematical functions.

The resulting system of equations constitutes an abstract representation of the physical reality being studied and can take many complex forms that cannot be analytically resolved in most cases. A numerical model is therefore established through the simplification of mathematical equations by discretization so as to obtain a linear system of equations that are easier to resolve. There are two main methods to solve these: the finite element method (FEM) and the finite difference method (FDM).

In the finite element method, the discretization of the domain being studied is possible thanks to a mesh (often triangular), which makes it possible to cover the entirety of the domain. This mesh is an approximate redefinition of the domain's geometry, which makes up an approximation space that is required to solve the equation.

Each elementary mesh in the approximation space of the discretized domain constitutes a finite element. Simple functions are thus defined locally on each element, ensuring the transfer between the space dimensions of the modeled object and the local coordinates of the finite element.

The finite differences method is older than the FEM but is also based on the principle of discretization using a mesh. In the FDM, the meshes are often quadrilateral, which makes it easy to implement this method, especially with regard to adapting it to simple problems. While the FEM solves the system of partial differential equations (PDE) through interpolation functions, the FDM substitutes the differential operators in the PDE by simple finite differences, hence the name of this method.

The FDM makes it possible to resolve a system of relations (algebraic formulations) for a discrete problem (numerical scheme) at the level of each elementary mesh. The FDM is very often used to model geosynthetics and their interfaces.

Furthermore, it must be noted that regardless of the method used, continuous modeling requires the definition of the boundary conditions imposed on the surface of the domain being studied, which allows us to guarantee the uniqueness of the solutions.

One of the chief advantages of continuous modeling is that it allows these conditions to be taken into account in a systematic and simple way, regardless of the shape of the domain.

3.1.2.2. *Using continuous methods*

Finite modeling methods can be easily applied to the reproduction of the behavior of structures in quasi-static and dynamic cases, especially when studying interactions between soil and civil engineering structures. The use of a matrix-type resolution system also facilitates a possible coupling of mechanical, hydraulic, and thermal phenomena.

As with any other model, finite methods require the parameterization of the mechanical characteristics of the modeled materials, for which more or less complex elastoplastic behavioral laws have been defined. The parameterization of these laws is carried out with regard to the physical properties of an experimental material, which was characterized in laboratory tests. The procedure can become complex when defining certain parameters for behavioral laws used, which are not always easily derived from characterization tests.

Although these parameters could be taken from existing literature, we cannot always be certain that the modeled mechanisms have been introduced correctly.

At present, certain developments make it possible to take account of behaviors under large strains, for example, the possibility of opening "joint"-type elements to model a crack whose location is known.

3.1.3. *General information: discrete modeling*

Discrete methods model a material using a set of particles interacting at the level of their points of contact. Consequently, these methods are logically applicable to the modeling of granular materials such as soil and make it possible to reproduce many rheological phenomena related to granular rearrangement resulting from the displacement or rotation of particles and to the emerging of discontinuities and ruptures.

It is thus possible to take into account mechanisms such as shearing and load transfers, like the arching effect caused by the redistribution of loads in the soil above a discontinuity (conduit, cavity, rigid inclusion, etc.).

Furthermore, the discrete nature of the set of particles used to model the soil makes it possible to reproduce phenomena such as bulking, compacting and cracking (in the case of a cohesive material), while also allowing large displacements (for example, taking into account the collapse of a soil layer above a cavity).

3.1.3.1. *General principles underlying discrete modeling*

The discrete modeling of granular materials is based on a description of the discontinuous material at the granular level. The material is simulated by an assembly of particles (or discrete elements) in contact with one another, with each element conserving all of its degrees of freedom through displacement and rotation. A contact law is defined between discrete elements, taking into account the contact stiffness and the criteria for tensile rupture, for compression, or for shear forces at the microscopic level of the intergranular contacts.

There are two different approaches, and therefore two solution methods, that coexist: the contact dynamic (Jean and Moreau 1992; Jean 1999) where the particles are considered to be indeformable, and the molecular dynamic, referred to in the applications discussed in the following section.

The molecular dynamic considers that the elements are indeformable, but may interpenetrate to a small extent at the level of a contact. Cundall (1971) introduced the discrete element method (DEM), formulated based on molecular dynamics to study the problem of rock mechanics. This method was further developed (Cundall and Strack 1979) and it has been widely adopted within the field of civil engineering.

The possibility that particles in contact will interpenetrate in a DEM problem makes it possible to establish a law between the interpenetration distance and reaction force, and an explicit solution method has thus been established (Figure 3.1). All application types require the choice of a time step small enough to ensure the convergence of calculations.

When posed in this manner, the problem is resolved over time in an iterative manner (Figure 3.1) using a succession of contact detection phases, computing the interaction forces for the discrete elements in contact, and then computing the positions and velocities of the particles in the next time step. Damping models (viscous or local) may be used to facilitate convergence in quasi-static applications by restricting the propagation of an elastic wave.

In the computation codes, the discrete elements can take any shape (polyhedron, sphero-polyhedron, etc.). In general, spheres or an assembly of many spheres (called "clumps" or "clusters") are used, since they make it possible to optimize the contact detection procedures and therefore limit the computation time. The interactions between the particles are governed by the definition of simple-contact laws (Figure 3.2).

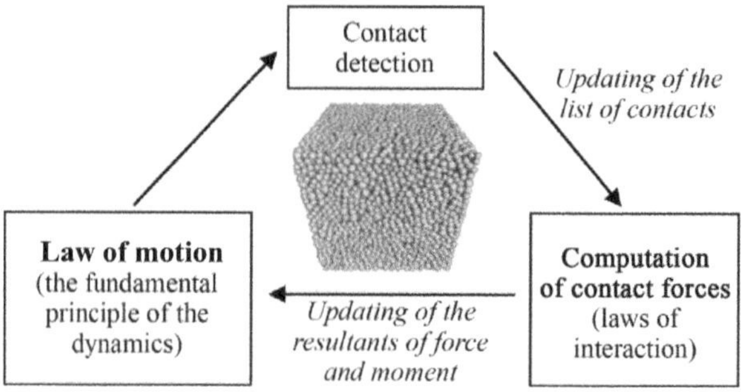

Figure 3.1. *Computation cycle in the DEM. For a color version of this figure, see www.iste.co.uk/villard/geosynthetics.zip*

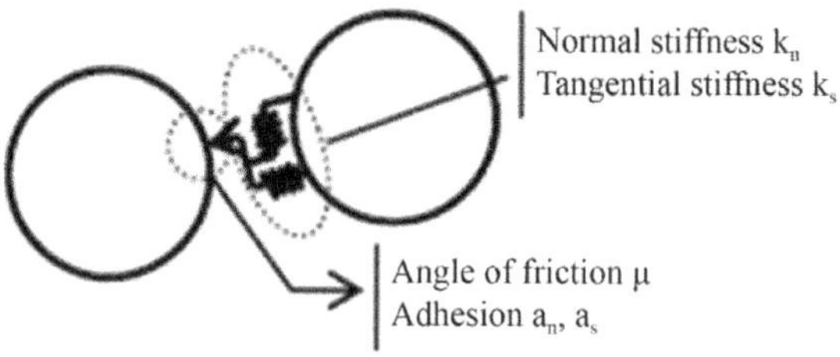

Figure 3.2. *Microscopic parameters of a contact between two discrete, spherical elements*

There are different ways of positioning the particles. For example, the Radius Expansion With Decrease Friction (REDF) method involving a gradual enlargement of spheres, allows the placement of elements with a given porosity, while guaranteeing precise control over their arrangement and their mechanical properties. "Pluviation" methods may also be used.

3.1.3.2. *Using discrete methods*

Discrete methods are easily applied to quasi-static modeling of granular materials, such as soils. Using discrete particles makes it possible to account for rheological phenomena that are not discernible when using continuous methods, for example, granular rearrangements, local discontinuities and ruptures resulting from the displacement and rotation of particles. It is also possible to take into account phenomena such as swelling and compaction in the case of a soil with high friction and low cohesion, or cracking in the case of a soil with significant cohesion.

These methods also offer the major advantage of making it possible to take into account large deformations in a granular assembly and to model, for example, sinkholes. It is also possible to model cyclical or dynamic phenomena.

The discrete nature of the modeling leads to the sample being generated by the increase in particle size until they reach the desired porosity, and then carrying out the actual modeling of the test on this sample. This first step of generating the sample is specific to discrete methods and may require considerable time, depending on the number of particles and the target porosity to be achieved.

Using discrete methods also requires calibrating the contact parameters between the particles at a microscopic level in order to obtain the desired mechanical characteristics on the macroscopic scale. This stage of the study results in a parametric study of the numerical characterization of the granular material's mechanical behavior and notably includes numerical modeling of triaxial tests, flexion tests, tensile strength tests, etc.

3.2. Applying continuous modeling

The continuous modeling of geosynthetic systems interacting with natural and anthropic materials (waste) dates back to the 1990s, with notable work being carried out by Wilson-Fahmy and Koerner (1993), Byrne (1994), Long et al. (1995), among others.

The modeling techniques have gradually evolved toward more complex techniques that make it possible to obtain more realistic results. A notable example are modeling techniques based on the FDM (section 3.1.2.1), which were developed by Fowmes et al. (2008), Sia and Dixon (2012), and Zamara et al. (2014). These techniques made it possible to take into account a certain number of key criteria related to the mechanical behavior of geosynthetic systems. More recently, inspired by earlier studies, a rational method of modeling geosynthetic systems was developed (Méthode Rationnelle de Modélisation des Géosynthétiques (MeRaMoG) (Tano 2016)). This method made it possible to model the mechanical behavior of geosynthetic systems and their interfaces in greater detail. This method enables the modeling of the actual number of geosynthetics and interfaces in the structure, taking into account the softening behavior of the interfaces and how the geosynthetic's stiffness evolves with deformation.

In order to do this, we must plot the curves showing how the stiffness evolves with the level of deformation of the geosynthetics, based on tensile tests (EN ISO 10319 (BSI 2015)) and EN 12311-2 (BSI 2013)), and also plot the curves showing how the angles of friction evolve with relative displacements to the interfaces based on direct shear tests or pull-out tests (EN ISO 12957-1 (BSI 2018) and EN ISO 13738 (BSI 2011)). These sets of evolving parameters, which define the behavioral laws of the geosynthetics and their interfaces, are then used in the computing software for the modeling. Subsequently, the tensile strength tests on the different geosynthetics and the shear tests carried out on the different interfaces are numerically modeled based on these variable parameters. The experimental and real results (tensile curves and shear curves) are then compared with the numerical results. Once this validation and calibration of the sets of parameters is carried out, it is possible to carry out numerical simulations across the entirety of the structure, including the geosynthetics.

3.2.1. *Principle behind the modeling of the geosynthetic reinforcement*

3.2.1.1. *Modeling the geosynthetic*

The mechanical behavior of the geosynthetics is directly influenced by their fibrous structure (geotextile and geotextile-related products) and the nature of the constituent polymers (polyethylene, polyester, etc.). It is generally accepted that geosynthetics exhibit an elastoplastic behavior, with an approximately linear elastic phase.

However, it must be noted that certain geosynthetics, such as HDPE membranes, exhibit strongly nonlinear behavior under stress. This is because the polymer's stiffness evolves depending on the level of deformation in the geosynthetic. In order to model the actual behavior of the geosynthetics, it is therefore essential to consider the evolution of the stiffness (Tano 2016).

Geosynthetics can also be subjected to extension, compression and bending. While their tensile behavior can be determined using standardized tests, at present, there is no method that enables the evaluation of their compression behavior. Consequently, compression phenomena are difficult to include in numerical modeling, as it is difficult to evaluate the stiffness of the geosynthetics under compression. With respect to the bending behavior, it is important to remember that geosynthetics have practically no resistance to the bending moments.

Furthermore, in continuous numeric models, the geosynthetics are modeled using so-called structural elements, in relation to thin structures with limited dimensions such as piles, anchors and reinforcement structures used in geotechnical engineering.

There are several structural elements whose behaviors (ability to deform, bend, etc.) are more or less different depending on the type of elements being considered (beam, cable, etc.). For example, beam-type elements are often used to model geosynthetics and possess a small bending strength depending on the moment of inertia attributed to them.

In all cases, for the most careful possible modeling of the geosynthetic behavior, it is necessary to use a structural element whose characteristics allow the following three conditions to be taken into account:

– being subjected to extensive tensile and compressive deformations resulting from tensile forces (Figures 3.3(a) and (b));

– being bent in order to simulate the membrane effect (Figure 3.3(c)). Considering that geosynthetics have very low bending stiffness and most operate in tension (Gourc 1982), the chosen structural element must only have very low bending strength;

– being displaced tangentially along the upper and lower surfaces of the chosen structural element (Figure 3.3(d)).

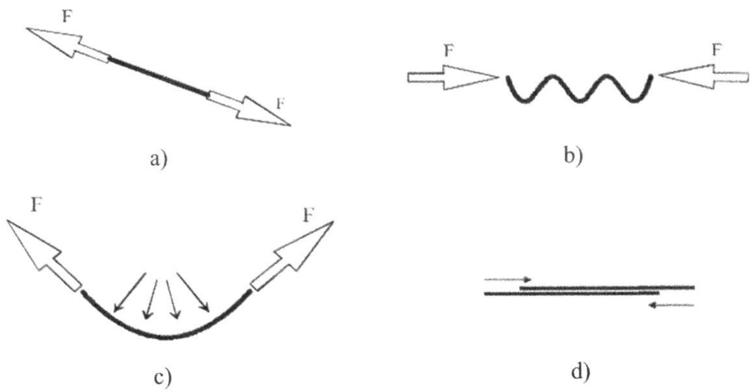

Figure 3.3. *Different kinds of behavior of a geosynthetic: (a) extension under stress; (b) folding through compression; (c) membrane effect under bending (Villard et al 2002); (d) tangential displacement through shear forces*

3.2.1.2. *Modeling the soil–geosynthetic interface*

Interface modeling makes it possible to take into account phenomena resulting from the "soil–structure" interaction, or "structure–structure" interaction, which are fundamental to working on geotechnical problems. These interactions, also called "contact problems", may involve the contact between a soil and a geosynthetic (for example, geotextile and underlying soil) or between two geosynthetics (for example, geotextile and geomembrane). In all cases, the forces transmitted to the geosynthetic are directly dependent on the mechanical behavior of its upper and lower interfaces.

To model these mechanical contacts, continuous modeling techniques make use of specific elements (for example, thin layer, hybrid method and zero thickness). Zero-thickness joint elements are the most widely used in calculation programs. As their name indicates, these zero thickness contact elements have no physical thickness within the model being developed. In the numerical resolution, a fictitious thickness is attributed to these elements. Using these elements thus makes it possible to ensure the continuity of the displacements and stresses at the level of the interface.

With respect to the mechanical behavior of interfaces between geosynthetics, it is generally accepted that they have a softening behavior characterized by the evolution of the shear strength with the relative displacement at the interface (Stark et al. 1996; Izgin and Wasti 1998; Flemming et al. 2006; Fowmes 2007; Eid 2011; Tano 2016). The friction forces mobilized increase gradually until they reach a maximum value (the peak) before decreasing and stabilizing at a lower value (residual). Consequently, considering only the residual characteristics may lead to an overestimation of the interface movements, while using the peak characteristics could lead to an underestimation of the displacements at the interfaces. Furthermore, using constant parameters could result in the modification of the distribution of shear stresses along the interfaces and, consequently, could result in a bias in the calculation of forces/deformations in the geosynthetic.

3.2.2. *Application of continuous modeling – Example 1*

3.2.2.1. *Description*

The first example of an application of this method is the modeling of an experimental device using MeRaMoG continuous modeling with the help of the finite difference software FLAC 2D. This experimental setup, which was developed to study the performance of reinforcement geosynthetics over cavities (Tano 2016), has a parallelpipedic shape with dimensions of 2 m × 1 m × 0.5 m (L × W × H) and a rectangular cavity measuring 50 cm at its center.

Figure 3.4 depicts the experimental setup. Here, the geosynthetic system tested is being compared to a Treatment, Storage, and Disposal Facility (TSDF) geocomposite barrier which, from top to bottom, is made up of a bentonite geosynthetic, an HDPE geomembrane and a non-woven PP geotextile. The entire barrier is supported by a PVA reinforcement geogrid, with a tensile strength of 50 kN/m. A bladder filled with water is used above the device to simulate the load of the waste on the safety barrier. Pressure is applied in stages of 10 kPa up to a maximum of 100 kPa (≈10 m of waste).

The deformations in the geogrid and those in the geomembrane are measured as the confining pressure increases (from 0 to 100 kPa) using pre-calibrated strain micro-gauges (10^{-6} ε) (Tano 2016).

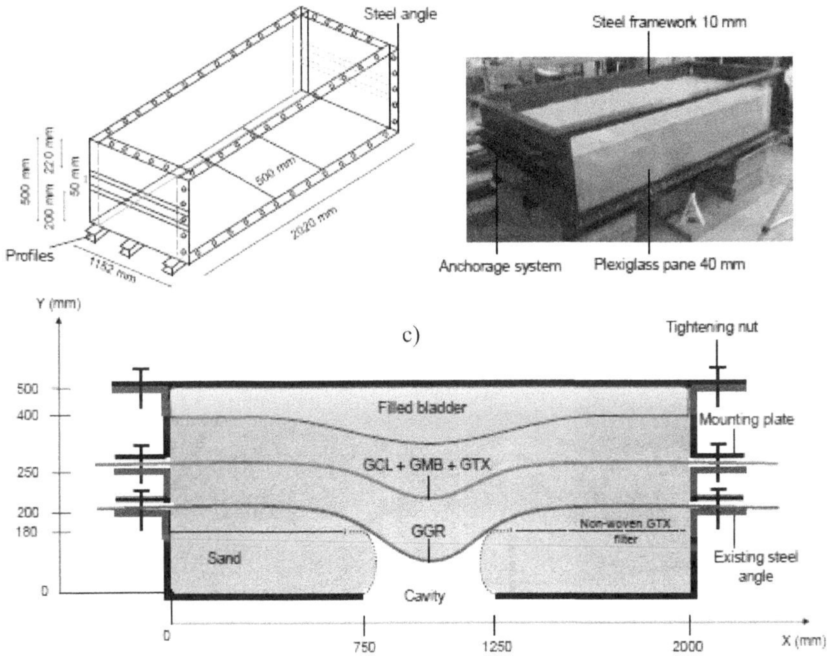

Figure 3.4. *Experimental setup: (a) dimensions; (b) photo; (c) position of the geosynthetics. For a color version of this figure, see www.iste.co.uk/villard/geosynthetics.zip*

3.2.2.2. *Configuring the numerical model*

Apart from the geogrid, all of the geosynthetics were modeled using beam-type structural elements without any bending inertia. These structural elements, defined in the software used, are the only ones that make it possible to represent a contact (interface) between two structural elements, and therefore between two geosynthetics. The strip element was used for the geogrid. This element was specially developed to model planar reinforcement structures, such as geogrids.

To take into account perfect anchoring conditions for the geosynthetics at the level of the lateral walls of the experimental device (Figure 3.4(c)), displacements of the nodes of the structural elements at the left and right extremities of the model were blocked. In addition, the elastoplastic Mohr–Coulomb model was used in association with the MeRaMoG method to model the mechanical behavior of the materials and interfaces. The MeRaMoG method makes it possible to take into account the softening behavior of the interfaces and the variation of the stiffness of the geosynthetics depending on the strain.

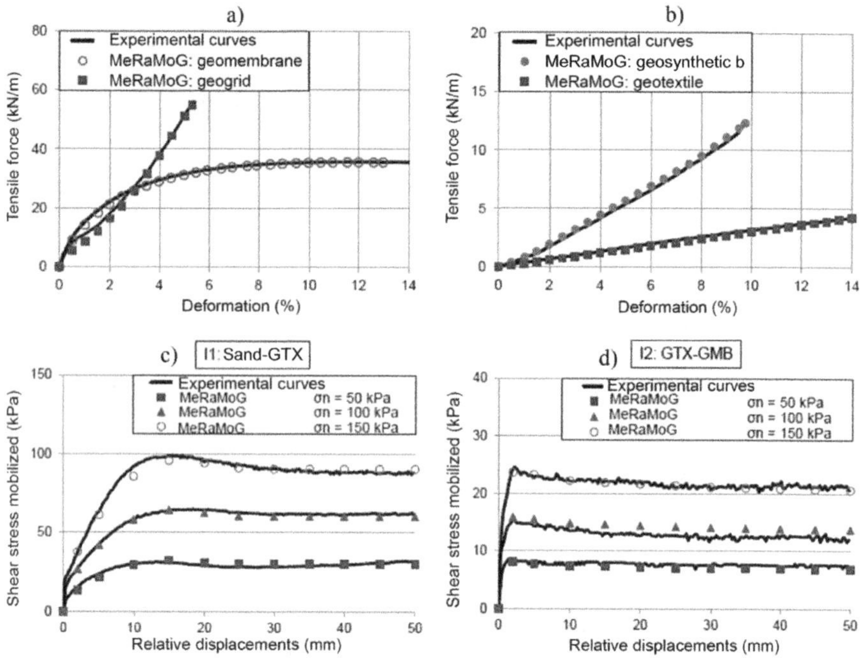

Figure 3.5. *Comparison between the experimental tensile strength and shear strength curves with those obtained through numerical simulation. For a color version of this figure, see www.iste.co.uk/villard/geosynthetics.zip*

COMMENT ON FIGURE 3.5.– *(a) Case of a smooth, 2 mm HDPE geomembrane and a PVA geogrid with a tensile strength of 50 kN/m. (b) Case of a bentonite geosynthetic and a non-woven PP geotextile. (c) Case of the interface between a sand and a non-woven PP geotextile. (d) Case of the interface between a non-woven PP geotextile and a smooth, 2 mm HDPE geomembrane.*

Tensile strength and shear strength tests were first carried out using the MeRaMoG on the four geosynthetics and the six interfaces, respectively. Next, the experimental results from these tests were compared with the numerical results using the sets of variable parameters (stiffness opposite the deformation of the geosynthetics, and the angle of friction opposite the displacement relative to the interfaces). The tensile strength curves of the geosynthetics as well as the shear strength curves of the interfaces are depicted in Figure 3.5.

An analysis of Figure 3.5 reveals an excellent agreement between the experimental data and the numerically simulated data. This agreement was verified for all geosynthetics and all interfaces, regardless of the level of deformation or displacement.

3.2.2.3. *Results*

Following the comparison of the experimental results of the shear strength tests at the interfaces and the tensile strength tests for the geosynthetics with the results obtained numerically, Figure 3.6 compares the experimental measurements obtained using the experimental setup with the values obtained in numerical calculations. These are the values for the deformation measured and calculated in the geogrid (Figure 3.6(a)) and in the geomembrane (Figure 3.6(b)) at various confining pressures.

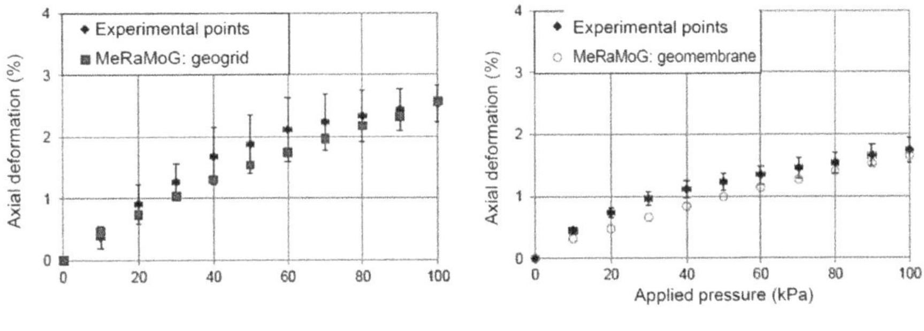

Figure 3.6. *Comparison between the experimental deformation measurements and the measurements obtained in numerical simulations using MeRaMoG: (a) PVA geogrid with tensile strength of 50 kN/m; (b) smooth, 2-mm HDPE geomembrane. For a color version of this figure, see www.iste.co.uk/villard/geosynthetics.zip*

It must be noted that this experimental setup simulates a loading combination consisting of bending forces, tensile forces, interactions between multiple geosynthetics, and high confining stresses (up to 100 kPa).

An analysis of Figure 3.6 shows that even in a case with complex loading, as described above, applying the MeRaMoG principles allows reasonable prediction of the deformations in the different layers of the geosynthetics. For example, when considering the geogrid, the simulated values are all within the margins of error of the experimental measurements.

3.2.3. *Application of continuous modeling – Example 2*

3.2.3.1. *Description*

This second example of an application relates to the study of the range of forces and strains developed within a geomembrane installed between two large piles of waste. This is a safety barrier in the typical case of the mixed extension of a TSDF, where the new storage unit is supported by the slope of the older storage unit. This safety barrier consists, from the bottom to the top, of a bentonite geosynthetic, a geomembrane and a protective geotextile. A 1-m-thick layer of soil combined with a foundation layer is modeled underneath this barrier. Finally, a reinforcement geogrid, installed within the foundation layer, was also considered.

The example given here relates to a comparison between the mechanical behavior of a smooth geomembrane and that of a textured geomembrane. The example presented here only considers the situation in which the new storage unit reaches a height of 20 m, the internal and external slopes of the units are 3H/1V and the residual angles of friction for the smooth and textured interfaces are 7° and 12°, respectively.

3.2.3.2. *Configuring the numerical model*

The mesh used for modeling the waste and soil comprises 6,400 volume elements whose size ranges from 1 m × 1 m to 2 m × 2 m. With respect to the geosynthetics, as in Example 1 above, beam-type elements were used to model the bentonite geosynthetic, the geomembrane and the geotextile. The strip element was also used for the geogrid.

With respect to the boundary conditions, the horizontal displacements of the mesh nodes were blocked on the left and right extremes of the model, since these points are far enough from the crest not to influence the results of the computations. All of the horizontal and vertical displacements were also blocked at the lower end of the model, considering that at this depth, the substrate is rigid enough to not be deformed.

Furthermore, in order to materialize the anchoring of the bentonite geosynthetic, the geomembrane and the geotextile at the level of the top of the existing storage unit, the horizontal displacements of the beam-type elements located at a distance X = 98 m were fixed. The geogrid was installed without any special condition in order to realize a flat anchoring linked to the load of the overlying materials.

Finally, the elastoplastic Mohr–Coulomb model, in association with the MeRaMoG method, was also used to model the mechanical behavior of the materials and the interfaces.

3.2.3.3. *Results*

Before discussing the results given in Figure 3.7, it must be stressed that these results are only valid for the chosen mechanical and geometric configurations. Figure 3.7 presents the results obtained for the smooth geomembrane (on the left) and the textured geomembrane (on the right) in terms of the shear stresses at the interface between the geotextile and the geomembrane (Figure 3.7(a)), the tangential displacements at the level of this interface (Figure 3.7(b)) and the forces and strains developed within the geomembranes (Figure 3.7(c)).

A quick analysis shows that, in general, the shear stresses that develop at the textured interface are higher (≈40% more) than those that develop at the smooth interface in the vicinity of the slope. It also shows the peak stress obtained for the case of the smooth geomembrane at the foot of the internal slope.

This peak is probably linked to a transfer of the non-mobilized shear stresses in the upper part of the slope, given the low strength characteristics at the interfaces (Ø = 7°). Further, this low shear strength results in large tangential displacements (1.5 m) as shown in Figure 3.7(b). On the contrary, the installation of a textured geomembrane allows the displacements to be limited to under 10 cm.

Finally, with respect to the forces and strains, Figure 3.7(c) indicates that on average, the textured geomembrane is able to tolerate five times the forces (concentrated at the head of the slope) in comparison to the smooth geomembrane (7.6/1.5). Nonetheless, it must be noted that the low shear strength of the interface with the smooth geomembrane leads to the transfer of shear stresses toward the foot of the slope, thus creating a significant local increase in tensile forces and strains.

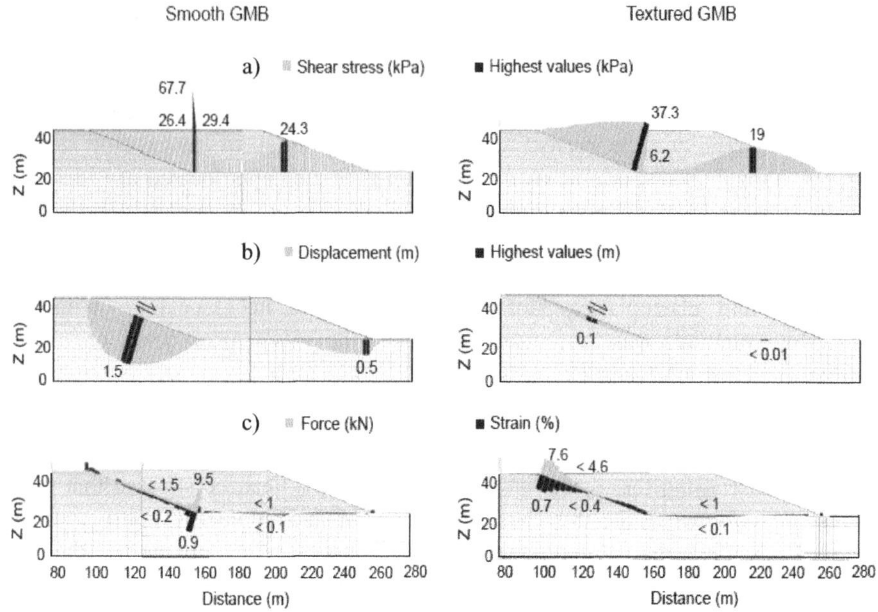

Figure 3.7. *Difference in the behaviors of a smooth geomembrane and a textured geomembrane: (a) shear stresses at the interface between the geotextile and the geomembrane; (b) tangential displacements at the interface between the geotextile and the geomembrane; (c) forces and strains within the geomembrane. For a color version of this figure, see www.iste.co.uk/villard/geosynthetics.zip*

3.3. Application of discrete modeling

Discrete methods have the advantage of modeling the kinematic behavior of the soil under large strains, as in the case of collapses, while using a relatively small number of parameters.

3.3.1. *Principle behind the modeling of the geosynthetic reinforcement*

3.3.1.1. *Approaches to modeling the geosynthetic*

Existing research shows two broad principles for modeling a geosynthetic reinforcement within a single 2D or 3D numerical model.

An initial approach consists of modeling the geosynthetic reinforcement using discrete elements such as spheres (Figure 3.8; Chen et al. 2012). The soil-reinforcement interaction is managed by the contacts between the discrete elements for the soil and for the geosynthetic. One of the drawbacks of using this method is the fact that despite very precise parameterization at the microscopic level, there is still some geometric roughness at the interface between the soil and the reinforcement.

It is also difficult to numerically reproduce the behavior of the reinforcement during a pull-out test, since the rigid link between the geosynthetic particles does not allow exact reproduction of its behavior (tensile, bending). Tran et al. (2013) also analyzed this model and concluded that the stresses and strains in the reinforcement could not be representative of reality.

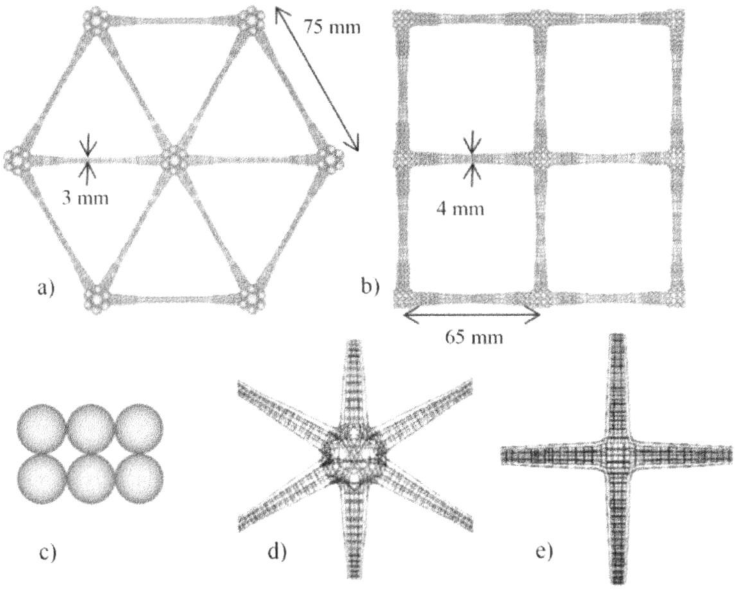

Figure 3.8. *Modeling a geogrid using the discrete elements method: (a) triaxial reinforcement; (b) biaxial reinforcement; (c) cross-section of the ribs; (d) connections over the triaxial reinforcements; (e) connections over the biaxial reinforcements (Chen et al. 2012)*

A second approach to overcome these drawbacks consists of a coupling between the discrete elements that represent the surrounding soil and the finite elements used for the geosynthetic reinforcement (Figure 3.9). Villard et al. (2009) therefore defined a model that was adapted to modeling reinforced soil structures. The finite elements that make up the geosynthetic are an assembly of plane triangular elements with three nodes, which take into account the structure of the geosynthetic: knitted, woven or nonwoven, reinforced in one or multiple directions. The behavior under tension and when acting as a reinforcement membrane is reproduced, even in the case of large strains. The interactions with the surrounding soil, discretized using discrete elements from the molecular dynamic (Cundall and Strack 1979), are achieved through specific interface laws, taking into account the rigidities for normal contact and tangential contact, and taking into account a contact friction coefficient.

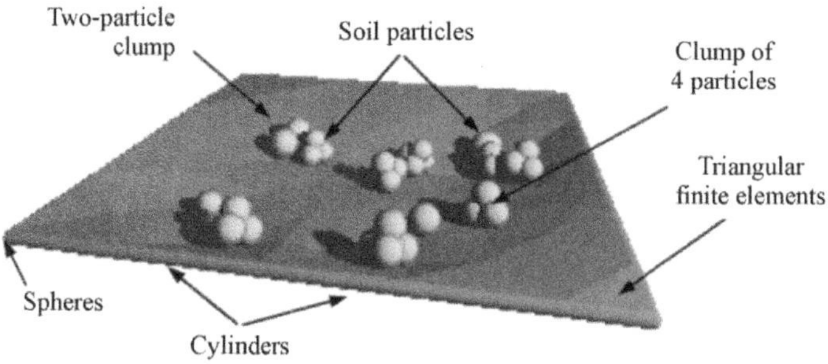

Figure 3.9. *Principle of the interaction between finite elements and discrete elements (Villard et al. 2009)*

Finally, there is a third approach consisting of coupling two computation codes of different types: one to model the reinforcement and the other to model the soil (Tran et al. 2013). It must be noted that managing the shift of data from one code to the other can result in a significant increase in computation time.

3.3.1.2. *Numerical characterization of the reinforcement in tension*

From here to the end of this chapter, applications related to the discrete methods will be approached using the numerical model developed by Villard et al. (2009). In this model, the geosynthetic reinforcement is characterized by its directions of reinforcement and the associated mechanical behavior law. The operator chooses a number of reinforcement directions and their orientation in the plane of the modeled geosynthetic. For example, a bidirectional, woven geosynthetic will be characterized by two perpendicular directions of reinforcement (Figure 3.10). A tensile behavior law for the geosynthetic fibers is then defined for each of these defined reinforcement directions. This law may be linear elastic through simplification, or it may be adjusted to the tensile strength curves resulting from tests on the geosynthetic.

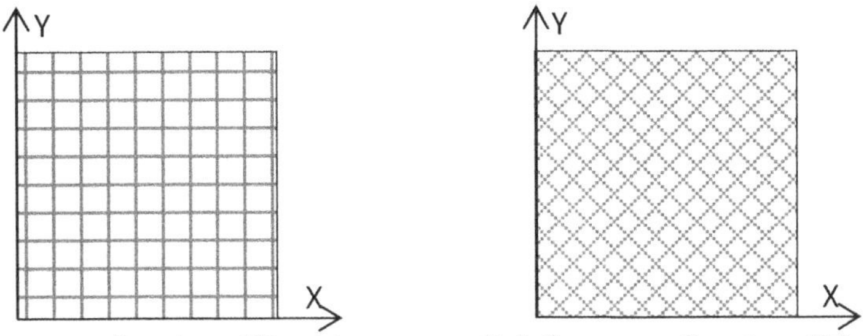

Reinforcement directions (/X-axis): $\alpha_1 = 0°$, $\alpha_2 = 90°$

Reinforcement directions (/X-axis): $\alpha_1 = 45°$, $\alpha_2 = 135°$

Figure 3.10. *Woven geosynthetic with two orthogonal reinforcement directions for two differently cut samples*

The numerical tensile behavior of the geosynthetic can be verified by simulating a tensile strength test on a strip of the geosynthetic (Figure 3.11). In the example depicted in Figure 3.11, the geosynthetic is characterized by two orthogonal directions of reinforcement. Each of these directions is associated with a linear elastic tensile behavior law with a stiffness J_1 = 10,000 kN/m and J_2 = 3,000 kN/m.

For the purposes of comparison, three tensile strength tests were simulated: a tensile test in the highest reinforcement direction for a sample measuring 0.2 × 0.1 m (test 1), and two tests with tensile forces at 45° to the two directions of reinforcement for samples measuring 0.2 × 0.1 m (test 2) and 0.2 × 0.2 m (test 3). The force is applied in the direction of the x-axis, and the edges of the geosynthetic are assumed to be maintained in place by rigid clamps.

It can be seen that the sample geometry (width of the sample and orientation of the reinforcing yarns) greatly influence the characterization result. The simplifying hypothesis consists of assuming that the reinforcement direction corresponds to the direction of the principal stress on the structure under study, and this thus brings up a limitation in the tensile characterization of the geosynthetic, especially in the case of structures subjected to uncontrolled stresses (for example, in sinkholes).

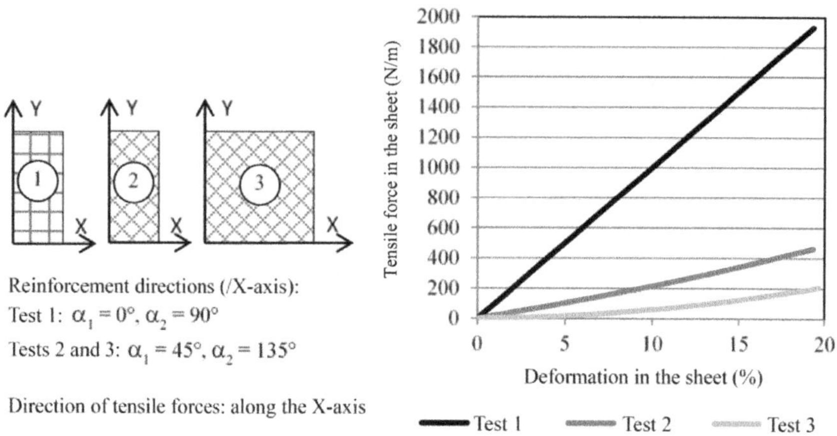

Figure 3.11. *Geosynthetic reinforcement characterized by three tensile strength tests*

3.3.1.3. *Numerical characterization of the soil–geosynthetic interface*

In the model proposed by Villard et al. (2009), the interface between the soil and the geosynthetic uses the parameters' normal stiffness, tangential stiffness and angle of friction.

The normal and tangential stiffnesses are similar to those used for the discrete elements and follow from the DEM. The normal stiffness is generally adjusted to the stiffness of the elements in the surrounding soil to avoid introducing discontinuities in the transmission of loads on either side of the geosynthetic.

Finally, the tangential stiffness must be carefully calibrated, especially with respect to the anchoring behavior of the geosynthetic: a value that is too small will result in very large tangential displacements and a value that is too high would involve a blockage of the reinforcement. These cases could result in unrealistic displacements and deformations in the geosynthetic and the surrounding soil, including regions outside of the anchorage zones.

Based on the value of the tangential stiffness and the normal stress on the geosynthetic, we can determine the value of the minimum relative displacement, U_0, required to mobilize the total friction given by the angle of friction between the soil and the geosynthetic. This angle of friction is generally obtained through pull-out tests or shear tests in the laboratory.

The minimum relative displacement for mobilizing the maximum friction at the soil–geosynthetic interface, introduced by Briançon and Villard (2006) is defined by a value of the order of a few millimeters. However, this parameter is not easy to quantify experimentally. This is another limitation of the numerical characterization of the geosynthetic and of its interface, regardless of whether the numerical modeling method is finite or discrete. Consequently, in discrete modeling, and for granular soil, carrying out a parametric study beforehand will permit refining of the choice of parameters in accordance with the reinforcement's experimental mechanical behavior.

3.3.2. *Example of application 1: wall made of reinforced soil*

3.3.2.1. *Presentation of the case under consideration*

Many structures require retaining walls (bridge abutments, slopes, etc.) and sometimes these must be built on soil with poor mechanical characteristics. Different construction methods make it possible to address the support issues and to minimize the weight of the wall. This is why Gorniak et al. (2015) studied the construction of a wall made of geosynthetic tubes containing expanded clay pellets (Figure 3.12(a)). More specifically, they studied the mechanical behavior of a 6-m-long tube under a cyclic load for different types of geosynthetic reinforcements using specific instrumentation. The basic reinforcement concept is that vertical loading on the tube will induce tensile forces in the geosynthetic and confinement of the clay pellets contained within.

a) b)

Figure 3.12. *(a) Construction of a retaining wall made of geosynthetic tubes (Gorniak et al. 2015); (b) loading test of a tube (Gorniak et al. 2016). For a color version of this figure, see www.iste.co.uk/villard/geosynthetics.zip*

The grading of the expanded clay aggregates varies from 10 to 20 mm, and the internal friction angle is estimated to be between 35° and 38° (Gorniak et al. 2016). The tubes were tested at two different fill percentages (85% and 100%, that is, a bulk density of 490–498 kg/m^3 and 540–560 kg/m^3, respectively) and different tube materials were tested: PP-PVA, PVA, PP and PES. The vertical surcharge is applied via a metal plate centered on the top of the tube, with static loads on top being increased until a total compressive stress of around 34.5 kN/m is reached (Figure 3.12(b)). The geometry of the cross-section of the tube and its height were measured using a 3D scanner after each loading stage.

3.3.2.2. *Discrete numerical modeling of the experiment*

The load is numerically applied on a cylindrical tube of unit length, with an initial radius R_0 = 0.3 m and containing 12,000 two-sphere clumps representing the expanded clay aggregates (Figure 3.13). These clumps involve the association of two interpenetrating spherical discrete elements that form a non-spherical, rigid particle and and they are placed at a porosity level determined using the Radius Expansion With Decrease Friction (REDF) method.

A total of five numerical granular materials (Soil-1 to Soil-5) are modeled with different initial porosities (0.34 – 0.36 – 0.38 – 0.4 and 0.42, respectively) to represent the different percentages to which the tube was filled. These granular materials have a Young's modulus between 11MPa and 29 MPa, and an internal angle of friction that varies between 28° and 46°. The interaction between the granular material and the geosynthetic is defined by an angle of friction of 30°.

Finally, the geosynthetic is modeled, in accordance with the experimental data, using the tensile stiffnesses of 59 kN/m in the radial direction and 43 kN/m in the longitudinal direction.

The numerical curves for the loading of the tubes (F_V), as a function of the height of the deformed tube (H), are compared in Figure 3.14 with the experimental values obtained at the end of the loading process. Since it is hard to control the fill rate for the tubes experimentally (two fill rates envisaged), five granular materials of different porosities were simulated for the purposes of comparison (Soil-1 to Soil-5).

In Figure 3.14, it can be seen that the experimental results produce two point-clouds characteristics of the two fill rates envisaged. It can also be seen that there is a good correlation between the two groups of tubes and the numerical results. Although we can only compare a few experimental points per tube, it is quite easy to identify the experiments that correspond to similar fill rates.

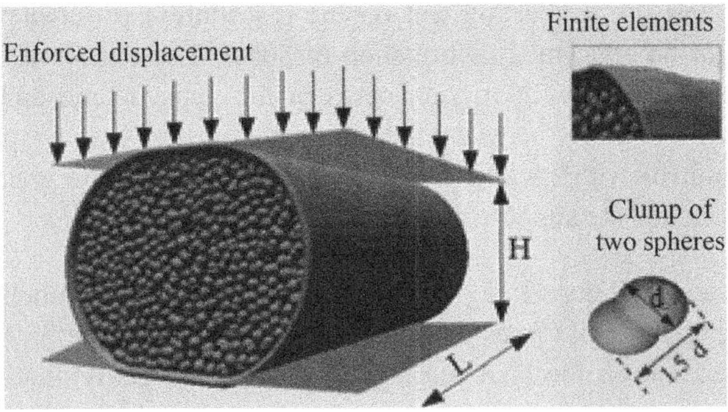

Figure 3.13. *Geometry of the discrete numerical model (Gorniak et al. 2013). For a color version of this figure, see www.iste.co.uk/villard/geosynthetics.zip*

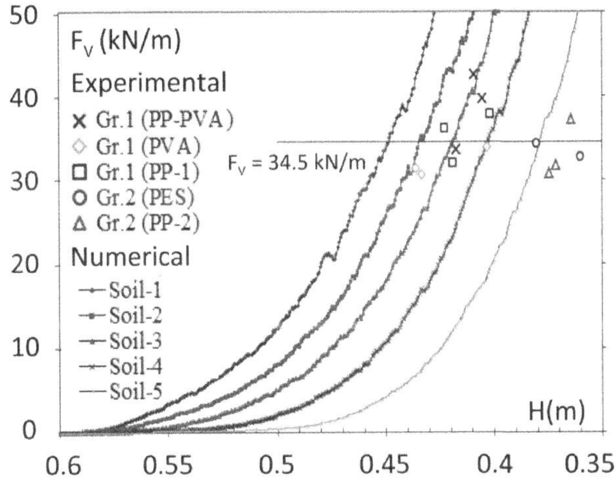

Figure 3.14. *Comparison of the experimental and numerical loading curves (Gorniak et al. 2013). For a color version of this figure, see www.iste.co.uk/villard/geosynthetics.zip*

One of the advantages of the numerical model is that it enables us to carry out systematic parametric studies in order to understand the role of certain parameters and to deduce general trends from this, which can then be used for design purposes. Among the results of the parametric studies carried out, Figure 3.15 illustrates the influence of the friction between the geosynthetic, the granular materials, the loading plates and the supporting soil on the distribution of tensile forces in the geosynthetic (for an axial tube deformation of 50%). The cases presented here are (a) without friction ; (b) with friction between the geosynthetic and the granular material; (c) with friction between the geosynthetic, the loading plates and the support soil; and (d) with friction between the geosynthetic, the granular material, the loading plates and the support soil.

As can be seen in Figure 3.15, the friction between the geosynthetic, the loading plates and the support soil (cases (c) and (d)) plays a crucial role in the distribution of tensile forces and on the load-bearing capacity of the geosynthetic tube, which increases. The cases without friction between the geosynthetic, the loading plates and the support soil (cases (a) and (b)) show a relatively homogenous distribution of tensile forces over the circumference of the tube. However, the influence of friction between the granular material and the geosynthetic is relatively small, as can be seen in the comparisons between cases (a) and (b) and cases (c) and (d).

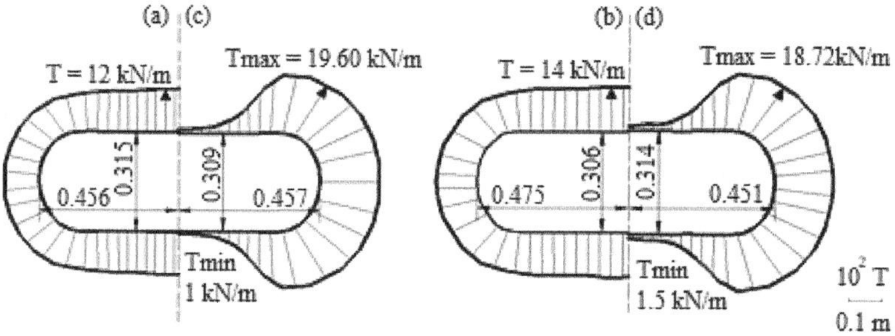

Figure 3.15. *Distribution of tensile forces over the circumference of the geosynthetic tube for the different hypotheses regarding friction (Gorniak et al. 2013)*

3.3.2.3. *Analytical deductions*

Since the numerical model was validated through true-scale values from experimental results, it has been used to carry out additional simulations which have made it possible to establish an analytical design method. The main hypotheses for this method are based on the hypotheses used for the numerical simulation: the tube is an elastic membrane of infinite length, without compressive or flexural rigidity; it only possesses tensile elasticity. The granular fill material possesses dilatancy and is frictional. The support soil and loading plate are considered to be non-frictional, and their self-weight is negligible compared to the applied forces. It can thus be derived that (Gorniak et al. 2015):

$$T = \frac{J\left(4R_0^{\,2}(1+\alpha)+H^2-4R_0 H\right)}{4\,H\,R_0} \qquad [3.1]$$

$$F_v = \frac{\pi\,J\left(4R_0^2(1+\alpha)-H^2\right)\left(4R_0^{\,2}(1+\alpha)+H^2-4R_0 H\right)}{8H^3 R_0 K_a} \qquad [3.2]$$

where T is the average tensile force in the geosynthetic, J is its elastic stiffness, F is the tube's load-bearing capacity, R is its initial radius, H is its deformed height, K_a is the coefficient of active earth pressure for the soil and α is the bulking coefficient of the soil under large strains.

Figure 3.16 depicts the comparison between the results from the numerical calculation and those from the analytic method. In this figure, it can be seen that the proposed analytic formula makes it possible to accurately approximate the compressive behavior of the geosynthetic tubes under stress from vertical loads.

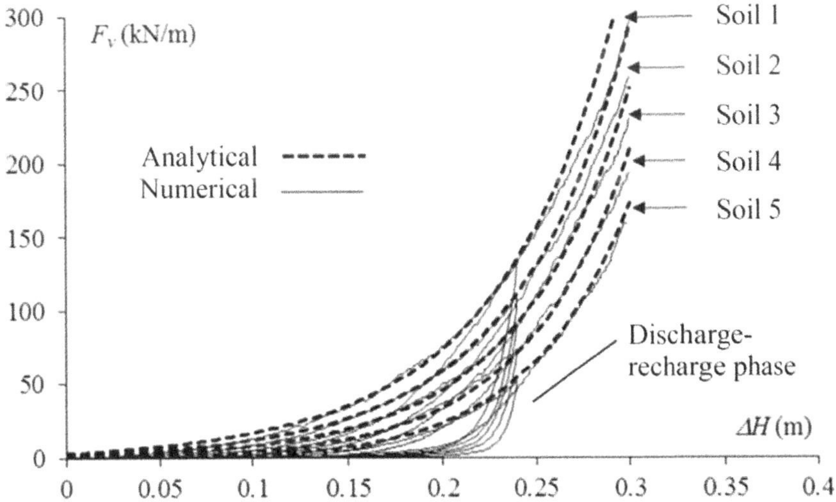

Figure 3.16. *Comparison of the numerical and analytical results (Gorniak et al. 2016)*

3.3.3. *Example of application 2: safeguarding against the risk of cavities*

3.3.3.1. *Presentation of the case under consideration*

The development of urbanization led to the creation of transport infrastructure over zones where the terrain had not yet been exploited, since they presented safety risks for the users. This is especially the case with zones at risk of collapse because of the presence of underground cavities, which cannot be detected in an exhaustive manner. Geosynthetic reinforcement at the base of a roadway or railroad platform offers a technical and economical solution that temporarily limits the consequences of a localized collapse toward the surface, while allowing trafficking to continue.

Although recent advances have led to improvements in the analytical design of geosynthetics over cavities (Briançon and Villard 2008), the mechanism underlying the transfer of the load within the granular layer and the vertical stresses acting on the reinforcement is still not well understood.

The GéoInov research project carried out full-scale experiments in order to understand the load transfer within the granular layer and the membrane behavior of the underlying geosynthetic reinforcement during the formation of a cavity (Huckert 2014). The diameter of this cavity increases gradually and reaches a value of 2.2 m.

The case considered here is an application using a PET geosynthetic and 1 m of round gravel backfill (20/40 mm). For every stage of the cavity opening and every loading stage, the instrumentation records the surface settlements, the displacements within the backfill, the deformations and sagging of the geosynthetic, and the vertical stress at the edge of the cavity.

3.3.3.2. *Discrete numerical study*

The numerical modeling of full-scale experiments is carried out using the model developed by Villard et al. (2009), which takes into account the large deformations related to the widening of the circular cavity (Figure 3.17). In accordance with the experiment, the model includes a 1-m-thick fill layer represented by discrete elements whose interaction parameters have been calibrated so as to obtain macroscopic characteristics for the fill material that are close to reality. That is, internal friction angles at rupture and residual friction angles of 36.5° and 31°, respectively, and a Young's modulus of 19 MPa. The woven structure of the geosynthetic is represented by two orthogonal directions of reinforcement with stiffnesses of 3,000 and 250 kN/m. The interfaces of the geosynthetic are characterized by a tangential stiffness of 15 MN/m, a friction angle of 23° with the overlying backfill, and a friction angle of 40° with the underlying soil. Two methods for the opening of a cavity are simulated: a gradual increase in its diameter up to 2.2 m ("sinkhole" process), and the lowering of a circular trapdoor with a diameter of 2.2 m ("trapdoor" procedure).

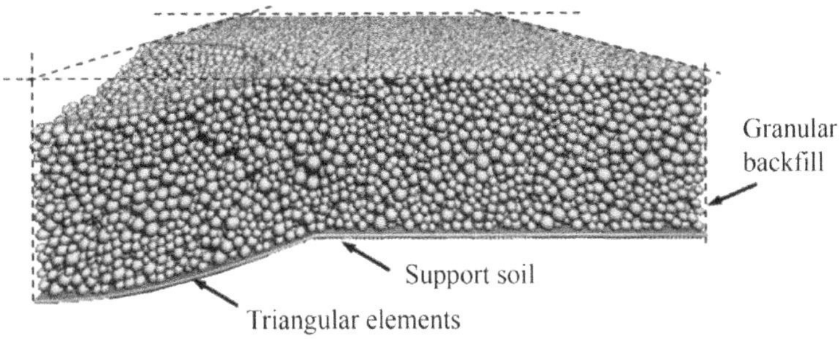

Figure 3.17. *Geometry of the numerical model (Huckert et al. 2014)*

The relevance of the numerical model was evaluated by comparison with the experimental results. Figure 3.18 gives the example of a comparison that proved to be satisfactory: a comparison between the surface settlements of the fill, experimentally and numerically, and the vertical displacements of the geosynthetic reinforcement in the case of a cavity being widened through an increase in its radius.

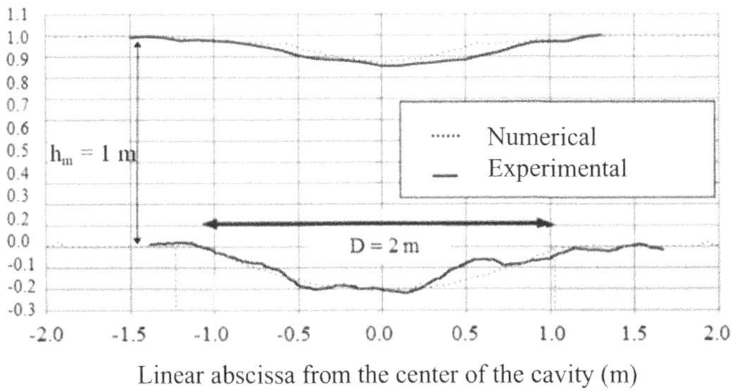

Figure 3.18. *Comparison of the experimental and numerical results of surface settlement and deflection of the geosynthetic (Villard et al. 2016)*

The discrete model allows an evaluation of the load distribution on the geosynthetic over the cavity for the two modes of opening that were tested (Figure 3.19). The vertical loading on the geosynthetic sheet is defined as the sum of vertical forces exerted by the discrete elements of the fill over a unit area of the sheet.

Due to the symmetry in the load transfer within the fill over the circular cavity, the reference zones for the computation are concentric rings centered on the cavity. It can be seen that the two modes of opening lead to very different kinds of loading on the geosynthetic reinforcement over the cavity.

The case of a sinkhole with a gradually increasing diameter leads to a conical distribution of the vertical stresses centered on the cavity, while in the case of a trapdoor, this distribution is inverted or almost constant.

In both cases, there is a significant concentration of forces at the edge of the cavity, testifying to the transfer of the load within the fill through an arching effect above the cavity.

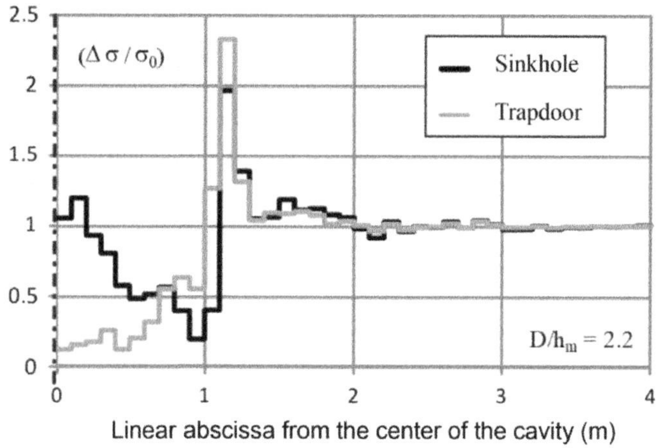

Figure 3.19. *Comparison of the distributions of vertical stresses over the geosynthetic for the two modes of opening of the cavity (Huckert et al. 2014)*

Thanks to the numerical model, it is possible to correlate this difference in the distribution of stresses on the geosynthetic reinforcement with other phenomena, such as the evolution of porosity (Figure 3.20) or the principal stresses (Figure 3.21) generated within the granular fill during the opening of the cavity. The development of the porosity of the fill is quantified by the ratio of the initial porosity to the final porosity at different points in the fill. In the case of a sinkhole with increasing diameter, the increase in porosity is more pronounced than in the trapdoor case, with the volume of sheared fill itself being greater due to the progressive increase in the diameter of the cavity and the consequent reorganization of the particles.

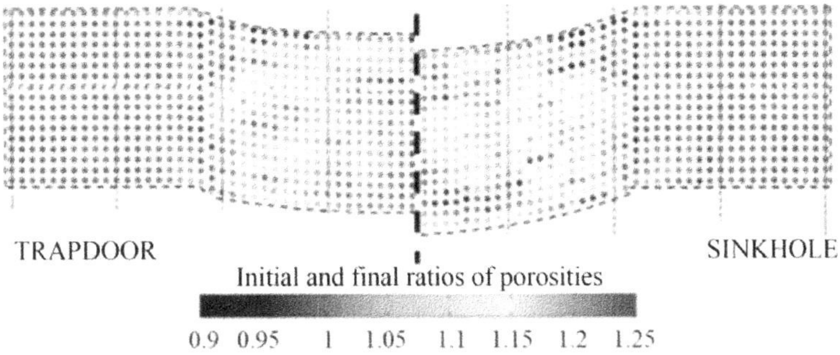

Figure 3.20. *Evolution of the porosity of the fill as a function of the mode of opening of the cavity (Villard et al. 2016). For a color version of this figure, see www.iste.co.uk/villard/geosynthetics.zip*

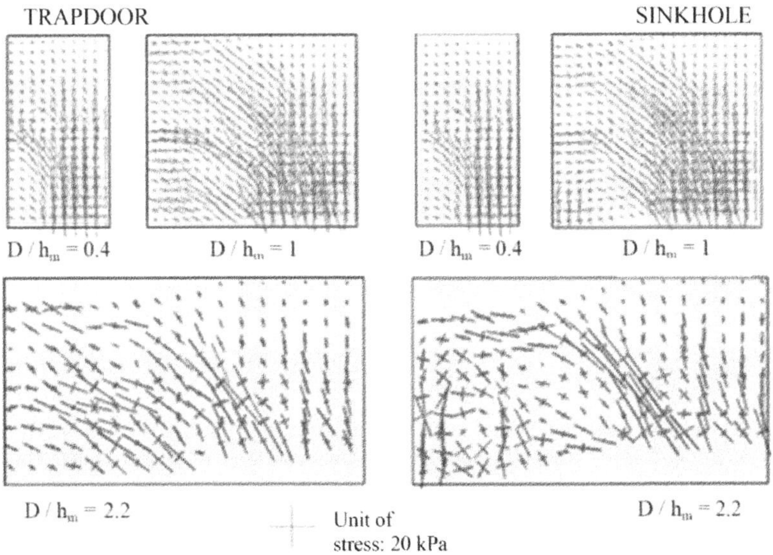

Figure 3.21. *Evolution of the distribution of principal stresses as a function of the mode of opening of the cavity (Villard et al. 2016). For a color version of this figure, see www.iste.co.uk/villard/geosynthetics.zip*

On the other hand, the study of the principal stresses during the opening of the cavity shows an increased concentration in the forces at the edge of the cavity for a trapdoor procedure, which corresponds to the highest peak in the stresses on the geosynthetic at the edge of the cavity (Figure 3.21). Furthermore, the case of a sinkhole with increasing diameter leads to the gradual destructuring of the fill above the cavity (connected with a higher porosity observed in Figure 3.20). The load transfers to the edge of the cavity are also modified throughout the formation of the cavity, which is not so in the case of a trapdoor (Huckert 2014).

As can be seen here, the discrete numerical model allows a detailed analysis of the mechanical functioning of the reinforced structure. The observation that the stresses are distributed non-uniformly over the geosynthetic has made it possible to optimize the existing analytical design methods by implementing new shapes of stress distribution over the reinforcement (Villard et al. 2016).

3.4. References

Briançon, L. and Villard, P. (2006). Dimensionnement des renforcements géosynthétiques de plateformes sur cavités. *Revue Française de Géotechnique*, 117, 51–61.

Briançon, L. and Villard, P. (2008). Design of geosynthetic-reinforced platforms spanning localized sinkholes. *Geotextile and Geomembrane*, 26(5), 416–428.

BSI (2011). EN ISO 13738, Geotextiles and geotextile-related products: Determination of pullout resistance in soil. Standard, BSI, London.

BSI (2013). EN 12311-2, Flexible sheets for waterproofing: Determination of tensile properties – Part 2: Plastic and rubber sheets for roof waterproofing. Standard, BSI, London.

BSI (2015). EN ISO 10319, Geosynthetics wide-width tensile test. Standard, BSI, London.

BSI (2018). EN ISO 12957-1, Geosynthetics: Determination of friction characteristics – Part 1: Direct shear test. Standard, BSI, London.

Byrne, R.J. (1994). Design issues with strain-softening interfaces in landfill liners. In *Proceedings of the Waste Technology 94*. American Society of Mechanical Engineers (ASME), Charleston.

Chen, C., McDowell, G.R., Thom, N.H. (2012). Discrete element modeling of the cyclic loads of geogrids-reinforced ballast under confined and unconfined conditions. *Geotextile & Geomembrane*, 35, 76–86.

Cundall, P.A. (1971). A computer model for simulating progressive large scale movements in blocky rock systems. In *Proceedings of the International Symposium on Rock Mechanics (ISRM)*, Nancy, 129–136.

Cundall, P.A. and Strack, O.D.L. (1979). A discrete numerical model for granular assemblies. *Geotechnique*, 29(1), 47–65.

Eid, H.T. (2011). Shear strength of geosynthetic composite systems for design of landfill liner and cover slopes. *Geotextiles and Geomembranes*, 29(3), 335–344.

Flemming, I.R., Sharma, J.S., Jogi, M.B. (2006). Shear strength of geomembrane-soil interface under saturated conditions. *Geotextile and Geomembrane*, 24(5), 274–284.

Fowmes, G.J. (2007). Analysis of steep sided landfill lining systems. PhD Thesis, University of Loughborough, Loughborough.

Fowmes, G.J., Dixon, N., Jones, D.R.V. (2008). Validation of a numerical modeling technique for multilayered geosynthetic landfill lining systems. *Geotextile and Geomembrane*, 26(2), 109–121.

Gorniak, J., Villard, P., Barral, C., Delmas, P. (2013). Analytical design method for predicting the vertical loading capacity of geosynthetic tubes filled with lightweight granular material. In *International Symposium on Design and Practice of Geosynthetic – Reinforced Soil Structures*. DEStech Publications, Inc., Bologna.

Gorniak, J., Villard, P., Barral, C., Delmas, P., Want, A. (2015). Experimental and analytical studies of geosynthetic tubes filled with expanded clay lightweight aggregate. *Geosynthetics International*, 22(3), 235–248.

Gorniak, J., Villard, P., Barral, C., Delmas, P. (2016). Coupled discrete and finite element modeling of geosynthetic tubes filled with granular materials. *Geosynthetics International*, 23(5), 362–380.

Gourc, J.P. (1982). Quelques aspects du comportement des géotextiles en mécanique des sols. PhD Thesis, Université scientifique médicale et Institut national polytechnique de Grenoble, Grenoble.

Huckert, A. (2014). Approches numérique et expérimentale du dimensionnement de renforcements géosynthétiques sur cavités et inclusions rigides. PhD Thesis, Université de Grenoble, Grenoble.

Huckert, A., Briançon, L., Villard, P. (2014). Experimental and numerical approaches of the design of geosynthetic reinforcements overlying voids. In *23rd European Young Geotechnical Engineers Conference EYGEC*. Warsaw University of Life Sciences (SGGW), Barcelona.

Izgin, M. and Wasti, Y. (1998). Geomembrane-sand interface frictional properties as determined by inclined board and shear box tests. *Geotextile and Geomembrane*, 66(4), 207–219.

Jean, M. (1999). The non-smooth contact dynamics method. *Computer Methods in Applied Mechanics Engineering*, 177, 235–257.

Jean, M. and Moreau, J.J. (1992). Unilaterality and granular friction in the dynamics of rigid body collections. In *Proceedings of the Contact Mechanics International Symposium*. Presses Polytechniques et Universitaires Romandes (PPUR), Lausanne.

Long, J.H., Gilbert, R.B., Daly, J.J. (1995). Effect of waste settlement on slope lining system. In *Proceedings of the Geosynthetics 95*. Industrial Fabrics Association International (IFAI), Nashville.

Sia, A. and Dixon, N. (2012). Numerical modeling of landfill lining system waste interaction: Implication of parameter variability. *Geosynthetics International*, 19(5), 393–408.

Stark, T.D., Williamson, T.A., Eid, H.T. (1996). Geomembrane/geotextile interface shear strength. *Journal of Geotechnical Engineering*, 122(3), 197–203.

Tano, B.F. (2016). Étude analytique, numérique et expérimentale du comportement mécanique des systèmes géosynthétiques : application au cas des barrières de sécurité des extensions d'ISDN. PhD Thesis, Université Grenoble Alpes, Grenoble.

Tran, V.H.D., Meguid, M., Chouinard, L.E. (2013). A finite-discrete element framework for the 3D modeling of geogrid-soil interaction under pullout loading conditions. *Geotextile and Geomembrane*, 37(1), 1–9.

Villard, P. and Briançon, L. (2008). Design of geosynthetic reinforcements for platforms subjected to localized sinkholes. *Canadian Geotechnical Journal*, 45(2), 196–209.

Villard, P., Kotake, N., Otani, J. (2002). Modeling of reinforced soil in finite element analysis. In *Proceedings of the 7th International Conference on Geosynthetics*, Nice.

Villard, P., Chevalier, B., Le Hello, B., Combe, G. (2009). Coupling between finite and discrete element methods for the modeling of earth structures reinforced by geosynthetics. *Computers and Geotechnics*, 36, 709–717.

Villard, P., Huckert, A., Briançon, L. (2016). Load transfer mechanisms in geotextile reinforced embankments overlying voids: Numerical approach and design. *Geotextiles and Geomembranes*, 44(3), 381–395.

Wilson-Fahmy, R.F. and Koerner, R.M. (1993). Finite element analysis of stability of cover soil on geomembrane lined slope. In *Proceedings of the Geosynthetics 93*, Vancouver.

Zamara, K.A., Dixon, N., Fowmes, G., Jones, D.R.V., Zhang, B. (2014). Landfill side slope lining system performance: A comparison of field measurements and numerical modeling analysis. *Geotextiles and Geomembranes*, 42(3), 224–235.

4

Barrier and Drainage Functions: Non-hazardous Waste Landfills and Hydraulic Structures

Nathalie TOUZE

SDAR, INRAE, Université Paris-Saclay, Jouy-en-Josas, France

4.1. Introduction

Non-hazardous waste landfills and hydraulic structures are both structures in which sealing must be ensured. In the former, this sealing is essential to protect the environment, while in the latter it conserves water resources.

The primary objective of this chapter, therefore, is to define the barrier function and the geosynthetic materials that make it possible to fulfil this function. Using these, it is possible to achieve a certain level of sealing.

While geosynthetics may have a secondary function, in addition to the primary function in other use-cases, this is never the case for the barrier function, which is the only function performed by a geosynthetic barrier.

Consequently, within a geosynthetic sealing system, the geosynthetic barrier is used in combination with other materials that fulfill the functions of protection, reinforcement, and drainage, for example.

For a color version of all figures in this chapter, see www.iste.co.uk/villard.geosynthetics.zip.

Geosynthetics from Yesterday to Today,
coordinated by Pascal VILLARD. © ISTE Ltd 2025.

In this chapter, we will study the drainage function provided by the drainage materials associated with the geosynthetic barrier within the geosynthetic sealing system.

We will therefore use reference documents (articles, regulatory texts, standards and technical guides) with the goal of explaining what the barriers are, looking at the materials that constitute the barriers, and also examining the restrictions on their usage in the specific cases of non-hazardous waste landfills and hydraulic structures. With respect to the hydraulic structures, the reference document used is Booklet 10 published by the *Comité français des géosynthétiques* (French Chapter of the International Geosynthetics Society). These are limited to dams, basins and canals.

In this chapter, we will first look at geosynthetic barriers. We then move on to geosynthetic sealing systems, and then the drainage function. In each section, we first present the general overview and then give specific information regarding the two contexts we are examining (non-hazardous waste landfills and hydraulic structures). We then move onto the question of equivalence, both for barriers and drainage structures. Finally, we discuss how defects in the geomembrane affect transfers within the barriers. A summary of key points is presented at the end of the chapter.

4.2. Geosynthetic barriers

4.2.1. *The barrier function*

The barrier function consists of preventing or limiting the migration of fluids. Geosynthetic barriers are materials that fulfill this function. They are defined in the standard NF EN ISO 10318–1 (AFNOR 2015) as a low-permeability geosynthetic material used in geotechnical and civil engineering applications to reduce or prevent the flow of fluid through a system.

4.2.2. *The geosynthetic materials used to make barriers*

4.2.2.1. *The different materials*

Geosynthetic barriers are classified into three sub-families, depending on the nature of the material performing the barrier function:

– clay geosynthetic barriers, whose barrier function is implemented by clays;

– bituminous geosynthetic barriers, whose barrier function is implemented by bitumen;

– polymeric geosynthetic barriers, whose barrier function is implemented by a polymer.

Clay geosynthetic barriers are commonly called geosynthetic clay liners (GCLs), while bituminous and polymeric geosynthetic barriers are commonly called geomembranes. Both families are more precisely defined in the following sections.

4.2.2.2. *Geomembranes*

4.2.2.2.1. General overview

Geomembranes have been defined in standardized definitions, given in the standard NF P84-500 (AFNOR 2013). Geomembranes are thin and flexible products, whose functional thickness is greater than or equal to 1 mm. Products with a functional thickness smaller than 1 mm are called geofilms. The minimum sealing level assured by a geomembrane, defined in standard NF P84-500, and measured in accordance with the standard NF EN 14150 (AFNOR 2019a), is 10^{-5} $m^3/m^2/j$.

Geomembranes are manufactured and transported to a site in the form of rolls of variable width (starting from 1.5 m). These rolls can have an area of up to 1,000 m^2 (or more). These rolls are watertight at the end of the manufacturing process. The continuity between rolls is ensured out by high-resistance, watertight assembly (Comité français des géosynthétiques 2017).

Geomembranes may have more or less smooth faces or rough surfaces. In the latter case, they are said to be textured geomembranes. The geomembranes may be equipped with and/or combined with one or more other components (composite geomembranes) (Comité français des géosynthétiques 2017).

Various polymers are used to manufacture geomembranes: high-density polyethylene (HDPE), polypropylene (PP), low-density polyethylene (LDPE), plasticized polyvinylchloride (PVC-P) and ethylene propylene diene terpolymer (EPDM) (Touze-Foltz 2010). Furthermore, additives (chemical components) are used in the manufacturing process to ensure the durability of the polymeric materials. The chemical and mechanical properties of the geosynthetics are highly dependent on the type of polymer used, the formulation of the additives, the shape of the geosynthetic and the application in which it is used (Hsuan et al. 2008).

4.2.2.2.2. Specific information for waste storage facilities

In the case of landfills, we differentiate between cases where the material is used for the bottom or internal slopes of the facility and cases where it is used as the cover. In France, the regulations related to non-hazardous waste landfills are specified in a decree dated February 15, 2016, which states that the geomembrane must be able to resist mechanical, thermal and chemical stresses throughout its operational life and long-term monitoring[1] (Journal officiel de la République française 2016[2]).

Therefore, when used for the bottom and the internal slopes, chemical stability is prioritized, along with the highest possible resistance to diffusive transfers that may be caused by pollutants in the leachate (the liquid formed after water percolates through the waste body). Pollutants contained within the landfill are substances that were not originally present in the surrounding medium and therefore a transfer occurs due to the concentration gradient between the interior of the waste storage facility and the exterior. It has been shown that HDPE geomembranes have the best ability to limit this diffusive transfer (Touze-Foltz et al. 2015). Studies carried out on HDPE geomembranes that incorporate an ethylene vinyl alcohol (EVOH) layer show that they have a greater ability to limit the diffusive transfer of volatile organic compounds. However, at present, we lack data on other families of pollutants. Thus, conventionally an HDPE membrane is installed at the bottom of the waste storage facility. In France, the conventional thickness of this layer, allowing the easiest implementation and the best sealing, is 2 mm.

When used as a cover for the landfill, while there is still the need to limit the diffusion of biogas into the surrounding environment, the most important factor is the flexibility of the material. As the waste degrades, there will be settlements in the waste body and there could be differential settlements, which the geomembrane must accommodate.

4.2.2.2.3. Specific information for hydraulic structures

In the case of hydraulic structures, the geomembrane must not only be adapted to the liquid to be contained and the desired lifespan, but must also be adapted to the form the structure necessary. In effect, for a structure with small dimensions, we will prefer a flexible geomembrane whose flexibility makes it easy to install and helps it adapt better to the form of the structure. This will reduce the risk of tension in the geomembrane and the emergence of sealing problems, both during installation as well

1 Defined in the decree as comprising the post-operational period and period for monitoring the environment, the duration of which cannot be less than 15 years for units with a single type of waste, and not less than 25 years for other storage facilities.

2 Official Gazette of the Republic of France.

as over the structure's lifespan. Consequently, while economic considerations are of importance, they are not the determining factor when choosing the geomembrane.

4.2.2.3. *Geosynthetic clay liner*

4.2.2.3.1. Definition

GCLs are defined by the International Geosynthetics Society as follows: an assembly of geosynthetics (conventionally two geotextiles) and a soil with low hydraulic conductivity (clay), which takes the form of a sheet, used in the field of civil engineering. A GCL variant consists of multicomponent GCLs, which are GCLs onto which a film or geomembrane is fixed, thereby improving the sealing function and/or protecting the material's clayey core (Von Maubeuge et al. 2011). Definitions related to GCLs are given in the standard NF P84-700 (AFNOR 2019b). However, in view of how important the barrier function is, it is important to define what "bentonite" is.

4.2.2.3.2. Bentonite

Bentonite is defined as a clayey material, composed chiefly of smectites (60% at least) and accessory minerals such as quartz, calcite, mica, feldspar, pyrites and goethite (Comité Français des Géosynthétiques 2011).

There are three main types of bentonite:

– Sodium bentonite: here the chief constitutive mineral is a sodium smectite.

– Calcium bentonite: here the chief constitutive mineral is a calcium smectite.

– Activated or modified calcium bentonite: this is a calcium bentonite that has been chemically modified by an ion-exchange process (known as activation). This is generally done by adding 5–10% of sodium carbonate. The chief mineral therefore is a sodium smectite.

Regardless of whether they are natural, artificial or activated, bentonites are crushed after drying, leading to different granulometries (powder and granules) that are suited to different purposes. Other treatments, such as the use of polymer additives, may also enhance the chemical resistance properties of the bentonites.

Since the swelling properties of sodium bentonites, calcium bentonites and activated calcium bentonites are all different, they do not offer identical sealing performances.

4.2.2.3.3. Constraints on the use of GCLs

GCLs can only fulfil the barrier function after the bentonite has been hydrated in the confining structure. The confining and protection structure must therefore exert a minimum level of and a uniformly distributed stress in order to limit the swelling of the bentonite and thus obtain the required level of sealing.

When a GCL is used to reinforce the passive barrier in case an equivalent protection to a clay liner is the goal, as will be seen in section 4.5, the minimal stress on the GCL, which lies under the geomembrane, is delivered by the drainage layer, which can be a mineral or combination of geosynthetics and granular materials. When used at the bottom of a structure, it is preferable that confinement takes place in the weeks following the installation of the geosynthetic barrier, as discussed in section 4.3 (Comité français des géosynthétiques 2011).

According to the recommendation guide for designing and evaluating equivalent provisions for a passive barrier in a waste storage facility (equivalence guide) (Ministère de l'Écologie, de l'Énergie, du Développement durable et de l'Aménagement du territoire 2009), the swell index of the clay, measured in accordance with the standard NF P84-705 (AFNOR 2019c), must be greater than or equal to 24 mL/2 g. This precludes the use of calcium bentonite (swell index of between 6 and 12 mL/2 g). The minimum mass per unit area of the dry sodium bentonite (natural or activated calcium) must be equal to 5 kg/m^2. The corresponding hydraulic conductivity of the GCL, measured in accordance with the standard NF P84-705, must be smaller than 5×10^{-11} m/s under a confining stress of 10 kPa, smaller than 3×10^{-11} m/s under a confining stress of 100 kPa, and smaller than 10^{-11} m/s under a confining stress of 160 kPa. Finally, the percentage of calcium carbonates, for natural sodium bentonites as well as activated calcium bentonites, must be smaller than or equal to 5% by mass.

When used on the internal slopes of the structure, the final confinement pressure will be provided by the waste.

When used as a cover material, in order to maintain the GCL's performance over time, the following must be ensured:

– Protection of the GCL from dehydration phenomena: the confining layers must be designed such that this goal is achieved. A minimum thickness of 1 m of confinement is recommended. In general, the thicker the confining layer, the smaller the risks of failure. The confining structure is brought about by the superposition of layers that carry out complementary functions such as drainage, water retention, anti-root barrier and plant support, and erosion control.

– Prevention of aggressive actions that could damage the GCL (mechanical aggression, vandalism, etc.).

In all cases, the minimum stress on the GCL from the confining and protective structure must be studied when the structure is designed. More detailed information can be found in the Comité français des géosynthétiques (2011) booklet dedicated to the use of GCLs in landfills.

More recent feedback from experience indicates that a soil thickness of 1 m for the confining layer may not be sufficient. Zanzinger and Touze-Foltz (2009) provides detailed information on the cover layer on GCLs that are used as coverings on structures.

There has not yet been enough feedback from the use of GCLs in hydraulic structures for recommendations to be finalized. However, it is imperative that the GCL is confined and maintained at a sufficient depth to prevent hydration-dehydration cycles and freeze-thaw cycles.

4.2.3. *Performance of the barriers*

Since the barrier function is the sole function performed by geomembranes and GCLs, flow is an essential characteristic to measure.

The protocol in the standard NF EN 14150 (AFNOR 2019a) consists of applying a pressure difference of 100 kPa on opposite sides of the geomembrane.

The measurement of steady-state liquid flow in GCLs is carried out as per the standard NF P84-705 (AFNOR 2019c). The hydraulic conductivity value can be derived by applying Darcy's law, except in the case of multicomponent GCLs, which are not porous. In the latter case, the flow is measured by combining the apparatus from NF EN 14150 with the measurement grid from the standard NF P84-705 (Barral and Touze-Foltz 2015). At the time of this writing, there is no single standard that can qualify multicomponent GCLs.

Table 4.1 gives information on the performance levels in terms of the liquid flow across different materials. Since geomembranes and multicomponent GCLs are nonporous, Darcy's law cannot be used to characterize their permeability to liquids. The flow is only in one direction. The data from Table 4.1 show that the sealing level of geomembranes, multicomponent GCLs and GCLs is much higher than other materials traditionally used in civil engineering.

Material	Hydraulic Conductivity (m.s^{-1})	Thickness (m)	Flow (m^3 /m^2 /j)
Concrete	10^{-10} (1)	0.1	9.5×10^{-5}
Roller-compacted concrete	10^{-8} (1)	0.5	2.6×10^{-3}
Bituminous concrete	10^{-9} (1)	0.1	9.5×10^{-4}
Compacted clay	10^{-9} (1)	1	1.7×10^{-4}
GCL	10^{-11} (1)	0.01	8.7×10^{-5}
Multicomponent GCLs	Not applicable	0.01	$< 2 \times 10^{-5}$ (2)
Geomembranes	Not applicable	0.001	$< 10^{-6}$ (2)

(1) Giroud and Plusquellec (2017).

(2) Touze-Foltz et al. (2016).

Table 4.1. *Hydraulic and dimensional characteristics and the flow across different barrier materials, for a hydraulic head of one meter, in the case of porous materials. For geomembranes and multicomponent GCLs, a pressure of 100 kPa was applied (Touze-Foltz 2018)*

However, the geomembranes are only effective against advective transfers when they are undamaged. In section 4.6.1, we will discuss the importance of quality control during the installation and the impact of defects on transfers within composite barriers consisting of a geomembrane combined with a soil that has low hydraulic conductivity or with a GCL. Section 4.3 examines the importance of designing the geomembrane or the GCL in the context of a geomembrane or GCL sealing system in order to ensure the protection of the material that must carry out the barrier function.

4.3. Geosynthetic sealing systems

Like a GCL, a geomembrane's sole function is to act as a barrier. Given the various constraints discussed previously and the other criteria to take into account, the geomembrane must be integrated into a system made up of several structures, each with its own function, as defined in the following section: geosynthetic sealing systems (Figure 4.1).

Definitions related to geosynthetic sealing systems are given in standard NF P84-500. A geosynthetic sealing system comprises a support structure, one or more geosynthetic barriers and the confining and/or protective structure.

4.3.1. *Support structure*

This is made up of the formwork layer and the support layer for the geosynthetic barrier (GM, GCL or a combination of the two as in cases of composite barriers) with possible layers to drain away water and gases, and the lower protection layer. This is installed on the base of the formwork and under the geosynthetic barrier.

The support structure must be stable by itself, both during the construction of the structure as well as its in-service period.

4.3.1.1. *Functions*

The support structure is supposed to protect the barrier from mechanical aggression. The constitutive materials must themselves be non-puncturing, free of vegetation and must be chemically compatible with the geosynthetic barrier.

During the construction phase, the support structure has the role of enabling or facilitating the installation of the geosynthetic barrier.

Following this, when the structure is operational and in use, the support structure must distribute the transmitted efforts while only generating acceptable levels of deformations in the geosynthetic barrier (the induced deformations must not compromise the sealing performance of the geosynthetic barrier, whether in the short term or the long term).

When one or more drainage systems are integrated into the support system, these must make it possible to cancel out underpressure effects (water or gas). Draining away water can, in certain cases, make it possible to detect and eventually capture accidental infiltrations.

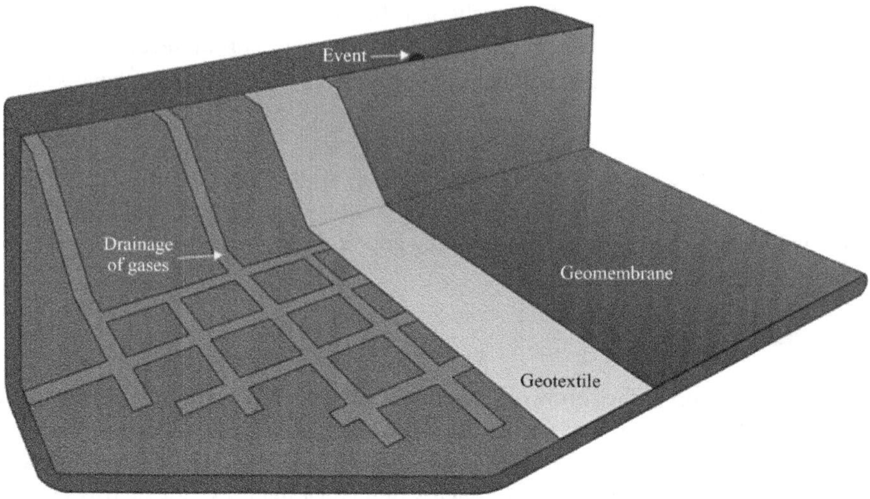

Figure 4.1. *Example of a geosynthetic sealing system (from Comité français des géosynthétiques 2017)*

4.3.1.2. *Materials for the formwork*

Depending on the structure and the site of the structure, the formwork can be made of a large variety of materials: clays, gravel, cement, etc. It must be approved by the site manager, the company in charge of the terracing or structure, and the company in charge of the geosynthetic installation.

4.3.1.3. *Materials for the support layer*

Depending on the structure and site of the structure, the support layer can be composed of a large variety of materials, whose function is to protect the geomembrane (geotextiles or geotextile-related products, sand, etc.).

4.3.1.3.1. Specific features of landfills

The decree dated February 15, 2016, concerning non-hazardous landfills (Journal officiel de la République française 2016) states that a landfill must be located on a site with favorable conditions in terms of geology, hydrology and hydrogeology. The sub-soil in the zone to be used is a passive barrier, which must not come under stress during operations on the site and must ensure long-term protection against the pollution of the soil and the groundwater and surface water by waste or leachate.

Regulations specify that at the bottom of the storage unit, the passive barrier must be made up, from top to bottom, of a layer whose hydraulic conductivity, K, is less than or equal to 10^{-9} m/s over a thickness of at least 1 m on top of a layer with hydraulic conductivity K, less than or equal to 10^{-6} m/s over a thickness of at least 5 m. On the sides, the hydraulic conductivity must be less than or equal to 10^{-9} m/s over a thickness of at least 1 m (Figure 4.2).

The "active" barrier is placed above this passive barrier and comprises a geomembrane that performs the barrier function and contributes to the drainage and collection of leachate and a drainage layer.

In the specific cases of projects that have a clay support layer, considered to be a passive safety barrier, we must avoid placing a drainage material between this support layer and the geomembrane (Comité français des géosynthétiques 2017). Thus, the use of a protective geotextile, preventing the geomembrane from getting dirty, and facilitating the seaming operation, is proscribed. It has been demonstrated that this geotextile will become a hydraulic conductor and if there is any defect in the geomembrane, it will lead to leachate flowing into the environment (Touze-Foltz et al. 2006).

4.3.1.3.2. Specific features of hydraulic structures

Unless the designer or the project manager state otherwise, a protective geotextile acting as a support layer must be placed between the formwork and the geomembrane. This geotextile must be selected by the project designer, who can use the Comité français des géosynthétiques Booklet "Recommendations for protection against damage of geomembranes" (this reference guide is being finalized at the time this book was published).

In certain cases, the placement of a thin geotextile (whose sole function is to maintain cleanliness) may prove necessary for placing and assembling the geomembrane in the appropriate conditions (Comité français des géosynthétiques 2017). This usage is permitted, contrary to regulations about such placement along the bottom or sides of landfills.

In the case of GCLs, an additional geosynthetic may be required to satisfy puncture protection (Comité français des géosynthétiques 1998).

4.3.2. *Upper protective structure*

4.3.2.1. *Functions*

An upper protective layer may be required atop the barrier material, within the geosynthetic sealing system, depending on how well the "support structure-barrier" complex responds to the various external stresses (without any loss in properties) to which it is subjected, both during the installation process as well as during its operational life.

In some cases, the protective structure may also play a role in enhancing the esthetic appearance of the structure or help it blend into the landscape. In such cases, technical qualities are not the only factors considered when choosing the material (Comité français des géosynthétiques 2017).

The upper protective structure must be self-stabilizing to prevent any more forces being added onto the geomembrane. However, placing any material atop the geomembrane does result in some additional stresses and we must therefore ensure stability against sliding and ensure the tensile strength of the various components of the geosynthetic sealing system. This inspection is carried out in accordance with the standard NF G38-067 (AFNOR 2017).

Based on the inspection result, it may be necessary to install a reinforcement structure to increase traction against the ground, as well as appropriate anchorings.

Unless the designer specifies otherwise, a transition layer (puncture-resistant geotextile, fine material, etc.) is placed between the geomembrane and the upper protective materials. This layer offers mechanical protection against aggressive elements, especially during installation (Comité français des géosynthétiques 2017).

4.3.2.2. *Specific information regarding landfills (bottom, internal slopes and cover)*

French regulations (Journal officiel de la République française 2016) specify that at the bottom of the landfill, the barrier must be covered by a drainage layer having a minimum thickness of 0.5 m. This layer is made up of a network of drains allowing leachates to be drained away toward a main collection pipe complemented by an artificial or natural granular structure, whose hydraulic conductivity, K, is greater than or equal to 10^{-4} m/s (Figure 4.2). This drainage layer must be resistant to mechanical, thermal and chemical stresses over the entire operational period and long-term tracking.

A puncture-resistant geotextile is placed between the geomembrane and the material that makes up the drainage layer, if the latter presents a risk of damaging the geomembrane (Journal officiel de la République française 2016).

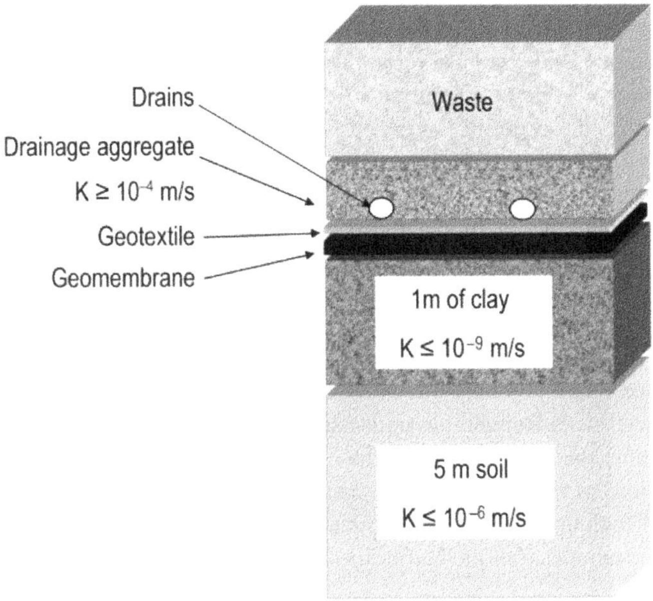

Figure 4.2. *Prescribed configuration for the passive and active barriers at the bottom of a landfill*

The barrier along the internal slopes of the landfill is covered by protective geotextiles or any equivalent material along its full height. In this context, it is possible to use drainage geocomposites. This arrangement is resistant to mechanical, thermal and chemical stresses throughout the operational period and the long-term usage.

When used for the cover, the geosynthetics may be used as part of the final cover which, from bottom to top, comprises the following (Journal officiel de la République française 2016):

– a barrier material;

– a runoff drainage layer composed of natural materials with a minimum thickness of 0.5 m or made up of geosynthetics;

– a covering soil layer with a minimum thickness of 1 m.

It must, however, be noted that this value of 1 m is not consensual as regards slopes, taking into account the risks of instabilities along the interfaces. When this layer is very thick, it poses difficulties with respect to performing the installation of the materials when there is a geosynthetic complex involved, which sometimes include a reinforcement geosynthetic as well as a drainage geocomposite. The reinforcement geosynthetic must resist the forces induced in the sealing layer due to the weight of the layers of natural materials lying atop it. Consequently, increasing the thickness of the covering layer results in an increase in the induced stresses, resulting in the need to use reinforcement geosynthetics with high tensile strengths. In certain cases, the range of required tensile strengths are at the outer limit or even beyond the capabilities of products that are currently on the market to fulfill this function. Furthermore, there are virtually no results from experience using these geosynthetics in this configuration (i.e. 1 m thickness of soil atop). Therefore, it does not seem reasonable to use them in this configuration.

The decree dated February 15, 2016, concerning non-hazardous waste landfills (Journal officiel de la République française 2016) also specifies that if the barrier is a geomembrane, the user must prove the implementation of best practices related to the installation of this material to ensure its effectiveness. Results from the inspection of each unit are kept available at the time of inspection of these classified facilities three months after the barrier has been installed.

Work on regrowing the plant cover begins as soon as the installation of the final covering layer is complete and this is done as per the modalities described in official orders from local authorities. The plants used are native, noninvasive species and they maintain the integrity of the barrier. This is notably achieved through plants whose root systems are compatible with the thickness of the soil layer and the future uses of the site.

4.3.2.3. *Specific information regarding hydraulic structures*

Various materials are commonly used to protect the geomembrane (Comité français des géosynthétiques 2017): natural materials (soil, sand, gravel, rip-rap, etc.), hydraulic cement that is poured on site, prefabricated concrete elements (slabs, interlocking tiles, etc.), or bituminous material.

The CFG's Booklet 10 (Comité français des géosynthétiques 2017) gives a detailed description of the stresses that must be taken into account when deciding on the requirement for, and the design of, a protective structure. This protective layer limits the aging of the geomembrane due to exposure to UV radiation, heat and oxygen, and also protects it from other climatic factors such as humidity and rain, as well from damage by animals, vandalism or the presence of floating objects.

It is also essential to have an accurate characterization of the soil to correctly gauge the durability of the nonexposed geosynthetics: pH, presence of oxygen, water content, organic matter, temperature and micro-organisms, the soil's carbonate content and its capillary water content (calcification must be prevented in the drainage layer) (Comité français des géosynthétiques 2017).

4.4. Drainage

4.4.1. *Definition of the concept of drainage*

The preceding sections illustrated the importance of the drainage function, whether below the geosynthetic barrier or above. The standard NF EN ISO 10318-1 defines the drainage function as the collection and transport of fluids in the plane of a geosynthetic. Thus, the role of the drainage layer is (Comité français des géosynthétiques 2014):

– to lower the groundwater levels;

– to control intermittent flows;

– to reduce and control water pressure;

– to accelerate consolidation phenomena.

The efficacy and durability of the drainage systems are ensured by combining the filtration and drainage functions, which requires the proper design of geosynthetics. The filtration function will be studied in Chapter 5.

Geotextiles and geotextile-related products, such as drainage geocomposites, can perform this drainage function.

A geotextile is defined as a permeable, planar textile material with a polymer base (synthetic or natural), which may be nonwoven, knitted or woven. Geotextile-related products are permeable, planar materials with a polymer base (synthetic or natural) used in contact with soil or other materials, in the field of geotechnology or civil engineering, and whose definition does not match the definition of a geotextile (AFNOR 2015). Drainage geocomposites come under geotextile-related products.

There are two fundamental aspects we must consider with respect to drainage (Comité français des géosynthétiques 2014):

– designing the in-plane flow capacity for the geosynthetic for a given distance to the collector;

– proving the stability of the structure when it is affected by the presence of the drainage provisions or the hypotheses related to its hydraulic design.

The in-plane flow in a geosynthetic is measured as per the standard NF EN ISO 12958 (AFNOR 2020). The value that is obtained is for a short-term measurement. The in-plane flow capacity is the volumetric outflow of water per unit width in the geosynthetic plane. This varies depending on the usage conditions of the geosynthetic, which are characterized by the hydraulic gradient of the flow, the compressive stress acting on the geosynthetic, the mode of application of this stress and the required usage period (Comité français des géosynthétiques 2014).

4.4.2. *Specific information related to waste storage facilities*

The decree dated February 15, 2016 concerning non-hazardous waste landfills (Journal officiel de la République française 2016) specifies that due to the prerequisites for the passive barriers, areas with thick alluvial soils are to be avoided when creating landfills. However, if it is not possible to avoid a region with thick alluvial soil, the impact study mentions technical steps that can be taken to prevent water being carried into the area that will be used. In such situations, the eventual drainage networks are to be installed such that they do not break the continuity of the passive barrier that is in place (Journal officiel de la République française 2016).

The drainage of leachates from the bottom, internal slopes and the cover is studied in section 4.3.

4.4.3. *Specific features of hydraulic structures*

4.4.3.1. *Provision for draining away water*

Water that may gather under the geosynthetic barrier can compromise the functioning of the structure (underpressures, thermos-osmosis, condensation, freeze-thaw cycles, erosion, loss of stability in the support or slope, etc.).

Unless the designer or site manager justifies any reason to the contrary, a "water drainage" structure must be provided for and adapted to the anticipated volumes of water. It is constructed using a granular layer, a drainage geosynthetic or drainage trenches.

Booklet 10 of the CFG (Comité français des géosynthétiques 2017) specifies that the granular layer must have hydraulic conductivity greater than 10^{-5} m/s and a minimum thickness of 0.1 m.

In some cases, a sandy layer that forms the drainage layer can play the role of a support layer.

The drainage geosynthetics have in-plane flow rates that have to be adapted to the structure. Their design can be carried out based on the CFG guide, "Recommendations for employing drainage and filtration geosynthetics" (Comité français des géosynthétiques 2014).

To prevent the clogging of the drainage material and to prevent entrainment of soil, a filter must be provided between the drainage layer and neighboring layers, in accordance with the state-of-the-art practices regarding filters for granular materials and geotextiles. Thus, in the case where a drainage system is to be installed under a clay geosynthetic barrier, we must ensure that the size of the openings in the draining materials, whether natural or geosynthetic, are adapted so as to prevent the risk of internal erosion of the clay through these openings.

The water that is drained away is gathered in collectors placed in the lower parts of the structure. If necessary, given the site of the structure or the volume of the outflows, a denser collector network is put in place.

In the ideal case where the drainage system makes it possible to detect and assess a leak, it must be verified that the outflow collected from the drainage system has not been increased by parasitic water inflows.

More details are given in the CFG booklets on geomembranes (Comité français des géosynthétiques 2017) and drainage (Comité français des géosynthétiques 2014).

4.4.3.2. *Provisions for gas drainage*

As with water drainage, appropriate gas drainage is required to dissipate gas underpressures (rapid rise of the water table into an unsaturated soil, the decomposition of organic matter at depth, leakage of liquids carrying organic matter, polluted soils, etc.). Unless justified otherwise by the designer, this system is systematically installed and must cover a minimum area of 20% of the structure's area. If the drainage is not carried out across the entire structure, it must be evenly distributed (Comité français des géosynthétiques 2017).

The drainage system must comprise:

– a drainage geosynthetic;

– a separation/filtration system that limits the clogging of the drainage system (for example, a filtration geotextile).

The gas drain outlets are installed in the open air at elevated points and must be protected (cap and grid) to prevent any obstruction and to prevent the entry of water or foreign bodies. The drain outlets and their connection must not allow runoff water to enter: the gas drainage does not work when wet. Therefore, unless there is a naturally draining support, a gas drainage system also requires the installation of a water drainage system.

4.5. Concept of equivalence

4.5.1. *Case of the passive barrier*

In the case of waste storage facilities, when the geological barrier does not naturally fulfil the conditions indicated in section 4.3.1.1.3, it is complemented and reinforced by other means that offer equivalent protection. The thickness of the barrier that is thus formed must not be smaller than 1 m when used at the bottom of the structure, and must not be smaller than 0.5 m for the internal slopes, up to a height of 2 m from the bottom.

All information related to the equivalence of the passive safety barrier is described in the application for permission for the facility. However, the regulatory text does not specify what "equivalent" means. Consequently, in 2002, the Ministry for Ecology and Sustainable Development created a working group to draw up a list of recommendations and to define a few best practice rules for applying the concept of equivalence. The institutes that were part of this working group were ADEME (French Agency for Ecological Transition), BRGM (French Geological Survey), Cemagref (a French public research institution for sustainable development in nonurban area, now INRAEs, now INRAE), CETE Lyon (Center for the Technical Study of Equipment), INERIS (National Institute for Industrial Environment and Risks) and INSA-Lyon (National Institute of Applied Sciences). This best practice guide was updated in 2009 (by the Ministry of Ecology, Energy and Sustainable Development, and Land use planning) and 2015.

GCLs can be used in this context of ensuring equivalency inside the passive barrier. The GCL is therefore brought in not to reduce the thickness of the passive barrier, but to remedy a permeability problem, in particular in the layer that is most permeable (Figure 4.3).

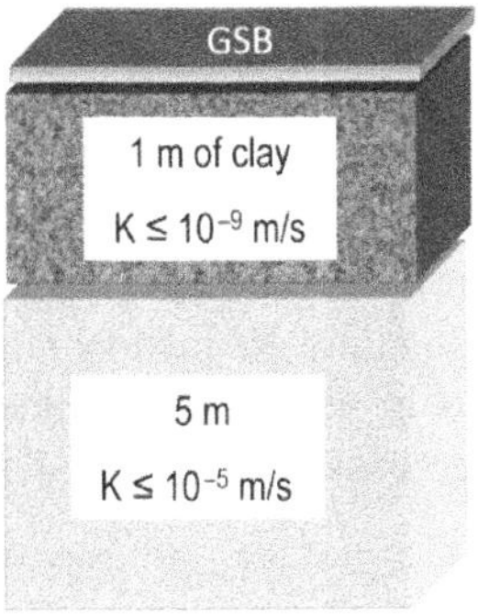

Figure 4.3. *Example of an equivalent configuration of the passive barrier at the bottom of the waste storage facility*

4.5.2. *Drainage*

The French regulations also open up the possibility of equivalence in the case of the active barrier, at least for the drainage layer, since at present there is no equivalent to a geomembrane. Therefore, the decree of February 15, 2016 concerning non-hazardous waste landfills (Journal officiel de la République française 2016) specifies that if an environmental risk assessment establishes that the landfills do not pose any potential risk to the soil, underground and surface water, or ambient air, then the requirements for the material for the drainage layer may be adapted through a Prefectoral order.

The system collecting leachate must be designed such that ideally, the maximum height of the leachates at the lowest point at the bottom of each unit is not more than 0.3 m above the geomembrane, while also not being greater than the thickness of the drainage layer. It must be ensured that this level can be monitored. Hence, the thickness of the granular drainage layer cannot be smaller than 0.3 m. The remaining 0.2 m can be replaced by a drainage composite, as long as the equivalence is demonstrated.

When used as a covering layer for the landfill, the provisions mentioned in section 4.3.1.2.2 may be modified by the Prefect if an application is made by the developer, as long as the planned construction methods guarantee that the effectiveness is equivalent to the effectiveness of the measures prescribed by the decree dated February 15, 2016. In any event, the sum of the thickness of the runoff drainage layer and that of the covering soil layer is greater than 0.8 m. The initial feedback from experience seems to show that installing this thickness is problematic and that it must be reduced.

Hydraulic equivalence cannot be assessed based on the in-plane capacity measured using the standard NF EN 12958. The value that is obtained is a short-term value. In reality, it will be impacted by different factors such as a decrease in the thickness of the geosynthetics over time due to compression creep (influenced by the temperature), the intrusion of the geotextile in the case of drainage geocomposites and clogging. Reduction factors whose values is greater than one must thus be applied to the initial value of the in-plane flow capacity. Thus, the flow value, q, that must be considered for the design is expressed in the form (Koerner 1999):

$$q = q_{12958}\left[\frac{1}{RF_{in} \times RF_{fl} \times RF_{cc} \times RF_{cb}}\right] \quad [4.1]$$

where:

– q_{12958}: the short-term flow measured using the standard NF EN 12958;

– RF_{in}: the reduction factor related to the filter intrusion;

– RF_{fl}: the reduction factor related to compression creep in the material;

– RF_{cc}: the reduction factor related to chemical clogging;

– RF_{cb}: the reduction factor related to biological clogging.

In the case of specific materials or specific configurations, other reduction factors may be envisaged (degradation related to installation, for instance). Although quite extensive, this list is not exhaustive but simply indicative of how these factors must be applied.

A working group from the International Organization for Standardization (ISO) is attempting to define the best methodology to evaluate in-plane flow for drainage geocomposites over the long term, especially in connection with the intrusion of the geotextile filter (Stoltz et al. 2017).

4.6. The impact of transfers within the barriers

4.6.1. *Importance of inspections during installation*

Legislators are well aware of the potential impacts of defects in the geomembrane. Thus, the decree dated February 15, 2016 regarding non-hazardous waste storage facilities (Journal officiel de la République française 2016) specifies that the geomembrane as well as the geomembrane assemblies, and therefore the seams, must be able to resist mechanical, thermal and chemical stresses over the entire operational period and long-term tracking, under normal operating conditions.

Furthermore, this version of the 2016 decree stipulates that the developer must ensure that the geomembrane is installed only by an installer who is certified in this field. The Comité français des géosynthétiques is involved in and supports the ASQUAL certification.

It is specified that the developer should call upon an independent, third-party organization to inspect the installation of the geomembrane. This third-party organizer ensures that the materials being installed have no manufacturing defect before installation on the site, and then inspects them after they are put in place.

A visual inspection of the geomembrane is carried out and complemented, a minima, by air pressure of the double seams with a central channel through which pressurized air is contained and by inspecting the single seams.

The controls mentioned above are carried out by a third-party organization. The developer puts in place a type of approval procedure for the installation of the geosynthetic. The results of the inspection are preserved on site and are made available at the time of the inspection of classified installations.

The CFG's Booklet 10 details the rules in force for the inspection and approval of the structure. There is also information regarding quality assurance (Comité français des géosynthétiques 2017).

4.6.2. *Quantification of transfers*

4.6.2.1. *Defects in geomembranes*

If all of the conditions are not fulfilled (the design of the geomembrane, the state of the appropriate support soil, certified installation, quality control throughout the installation, checks performed after installing the granular material, etc.), then the geomembrane may exhibit defects. The process does not reach an end point until the quality control of the geomembrane is complete, with no defects found (Comité français des géosynthétiques 1995, 2017).

There are different methods for detecting defects. Most nondestructive methods (which do not involve sampling from the material, especially at the seams, to carry out tests on the samples) are listed in the CFG guide on this subject (Comité français des géosynthétiques 2002), which is currently being revised. These methods can be divided into methods that can test the seams, those that can test the entirety of the geomembrane surface after installation, or even certain methods, like the electric methods, that can be used after the covering material is laid on top of the geomembrane. Table 4.2 summarizes the different methods used as well as their main limitations. A recent article (Charpentier and Jacquelin 2019) presents the two main methods used for the electric prospecting of geomembranes based on ASTM standards.

The water-jet technique is a geo-electric method that makes it possible to detect potential perforations that can be formed during the installation of the geomembrane (ASTM International 2003). This method uses the principle of the electric insulation of the geomembrane to locate defects. A constant electric voltage (DC) is applied through a metallic lance and an earthing unit is placed outside the site in order to stimulate a potential different between the water in the lance and the geomembrane foundation. Upon the detection of a defect that is large enough to allow water to pass through, a film of water crosses this defect and creates an electric bridge between the positive tension and the grounded earth unit. A sound signal is produced to warn the specialized technician about the presence of the defect. The water jet wets 100% of the surface area to be checked and thus validates the whole surface. This technique allows the detection of defects that are smaller than 1 mm^2.

Method	Geomembrane	Sheet	Seams	Covered geomembrane
Vacuum box	All	No	Yes	No
Screwdriver	All	No	Yes	No
Pressure in the central channel	Those joined using double seams	No	Yes	No
Ultrasound	Bituminous HDPE	No	Yes	No
Conductive geomembrane	Conductive geomembrane	Yes	Yes	No
Water jet	Nonconductive geomembrane	Yes	Yes	No
Dipole	Nonconductive geomembrane	Yes	Yes	Yes

Table 4.2. *Types of geomembranes that can be prospected at the level of the sheet and seams, with or without a covering material, using the main methods described in the guide (Comité français des géosynthétiques 2002)*

In cases where the geomembrane is covered, the dipole technique is recommended (ASTM International 2016; Charpentier and Jacquelin 2019). This method enables the detection of potential defects caused by the installation of the covering material. It also makes use of the insulation property of the geomembrane to locate defects. A continuous current is directly fed into the covering material and a grounded earth unit is placed outside the site so that the electric current must pass through any eventual defects in the geomembrane, thus generating a typical electric field that can be identified on the surface by a qualified technician.

In the case of ponds, for example alongside highways, for economic reasons we may choose to fill the pond with water to verify that it is watertight, although this option is questionable from an environmental point of view.

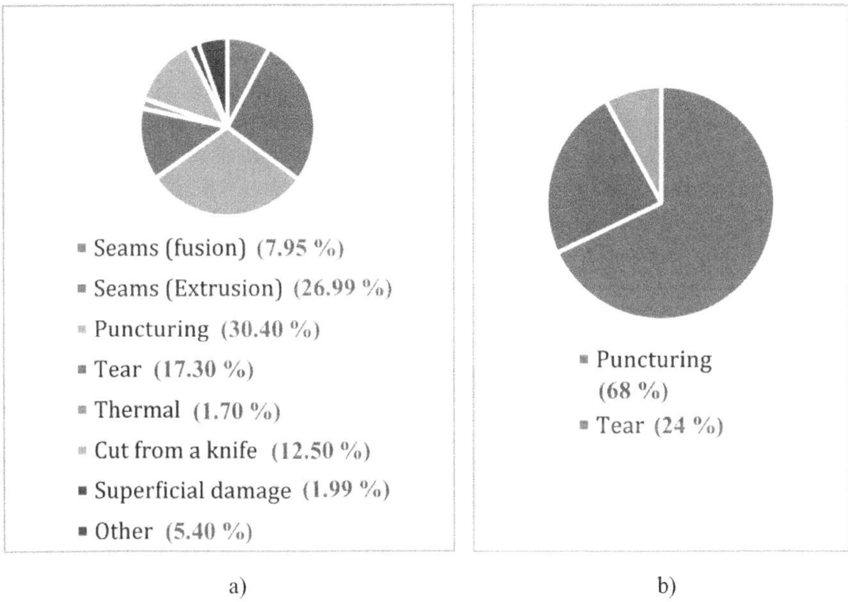

Figure 4.4. *Types of defects detected using (a) the water jet method and (b) the dipole*

Various articles, including Nozko and Touze-Foltz (2000) and the synthesis carried out by Touze-Foltz et al. (2008), list out the typology and the frequency of the defects. The most recent data are presented in Charpentier and Jacquelin (2019). Figure 4.4 presents the typology of the defects detected through the water jet methods and dipole methods, respectively, over 2.6 million m^2 of prospected geomembrane (Charpentier and Jacquelin 2019). The authors state that in 17 years, they have found defects in all but four of the prospected structures, which indicates that newly installed geomembranes frequently exhibit defects prior to inspection.

Based on the different geoelectric prospecting carried out, the defects can be classified into two categories: circular or longitudinal (Figure 4.5). Table 4.3 gives some information on the average density of defects per hectare. These values are consistent with the value of 12 defects per hectare proposed by Giroud and Bonaparte (1989).

With respect to the size of the defects that were detected, they are typically 1 or 2 mm^2, when detected using the water jet method. Once they are covered, this size increases. Thus, on average, 57% of the leaks detected using the dipole methods were smaller than 1 cm^2, 31% were between 1 and 10 cm^2, 16% between 10 and 100 cm^2 and 8% of the leaks measured over 100 cm^2, or 10 cm by 10 cm. These figures show that unless defects are detected, some structures do not offer the required level of sealing even immediately after construction (Charpentier and Jacquelin 2019). They also indicate the need to pay particular attention to protecting the geomembrane during both the design and the installation of the support layer, and the need to inspect the installation of the geomembrane. The CFG's Booklet 10 (Comité français des géosynthétiques 2017) may be consulted for information on quality control and quality assurance.

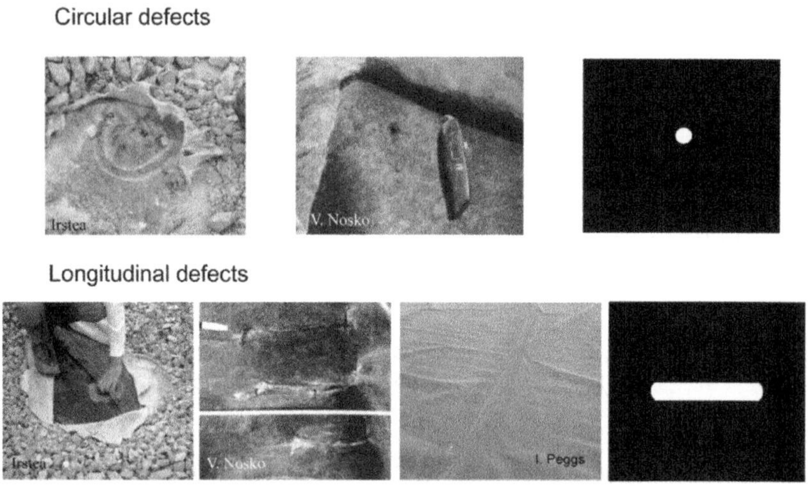

Figure 4.5. *Examples of circular defects and longitudinal defects and a conceptualization of both*

Reference	Prospected area (ha)	Defects in the geomembrane (%)	Seam defects (%)	Average density (/ha)
Laine and Mosley (1993)	1	20	80	8.3
Board and Laine (1995)	2	31	69	5.5
Colucci and Lavagnolo (1995)	25	85	15	15.3
White and Barker (1997)	1	100	0	0.7
Darilek and Miller (1998)	1	100	0	0.9
Snow et al. (1999)	2	100	0	10.9
Nosko and Touze-Foltz (2000)	325	93.7	6.3	12.9
Touze-Foltz (2001)	108.8	81.5	18.5	11.9

Table 4.3. *Statistics on the defects detected in covered geomembranes using geoelectric prospecting (from Touze-Foltz 2001)*

4.6.2.2. *Quantifying the impact of defects in the geomembrane*

Transfer within the barriers depends not only on the characteristics of the defect in the geomembrane, but also on the underlying medium, and on the quality of contact between the geomembrane and this medium. Consequently, if the contact is not perfect, the liquid that infiltrates through the defect in the geomembrane can flow between the geomembrane and the support soil before flowing into the soil (Figure 4.6).

According to Rowe (1998), there are three causes for imperfect contact between the geomembrane and the underlying medium:

– asperities on the soil surface related to the soil granulometry;

– undulations or ruts in soil surface, giving the appearance of a non-smooth surface;

– wrinkles in the geomembrane.

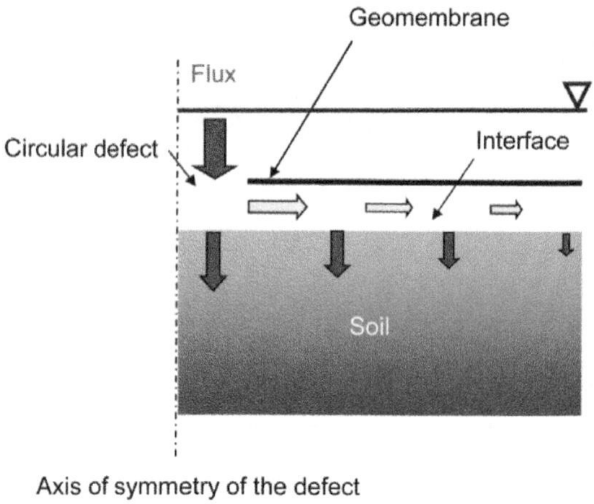

Figure 4.6. *Schematic diagram depicting the flow within a composite barrier in the presence of a circular defect in the geomembrane*

Thus, as mentioned previously, the quality of installation and the pre-installation inspection are key factors. Based on this, Giroud (1997) defined two contact conditions: good and poor, with the later addition of "excellent" contact (Touze-Foltz and Giroud 2003):

– Poor contact conditions (PCC) are those where a geomembrane is installed with a number of wrinkles and/or placed on a soil with low permeability, which has not been properly compacted and whose surface does not appear smooth.

– Good contact conditions (GCC) are those where a geomembrane is installed with as few wrinkles as possible, on a soil with low permeability that has been properly compacted and whose surface is smooth. Furthermore, it can be hypothesized that adequate confinement will allow the geomembrane to lie flat against the support soil.

– Excellent contact conditions (ECC) are those where a geomembrane free of wrinkles is installed over a clay geosynthetic or a soil with low permeability whose surface is exceptionally smooth and flat. Furthermore, it can be hypothesized that adequate confinement will allow the geomembrane to lie flat against the support soil.

In addition to the equations obtained for poor, good and excellent contact conditions, Touze-Foltz and Giroud (2003) and Bannour and Touze (2019) developed empirical equations for the specific cases of geomembrane-clay geosynthetic composites. This was done on the basis of laboratory experiments and the equations were for the case of virgin GCLs (hydraulic conductivities of the CGL lower than 10^{-10} m/s) and after aging through a cation exchange process (hydraulic conductivities greater than 10^{-10} m/s).

Different methods are used for these different contact conditions to estimate the rate of the leakage caused by defects in the geomembrane. In the most complex case, where the geomembrane is laid on a soil with low permeability, the simplest quantification is carried out using empirical equations. Considering the complexity of the soil surface state and the geomembrane's surface state, it does not seem realistic at present to create a 3D model that will take into account all of these nonuniformities. The complexity of the flow at the soil-geomembrane interface is shown in Figure 4.6, depicting the flows traced using a blue dye on the surface of clay and on a GCL, at the end of experiments carried out in situ and in the laboratory.

Figure 4.7. *Image of flow areas under a geomembrane, caused by a defect with a diameter of 4 mm in an HDPE geomembrane, after disassembling test sections of a clay geomembrane composite liner with a 7 m edge and a CGL-geomembrane composite liner with a diameter of 1 m*

Table 4.4 summarizes the main empirical equations that are recommended, based on the size of the circular defects, and Table 4.5 lists the equations for longitudinal defects (cuts and wrinkles).

The equations given in these tables were developed for, and can be applied to, the following parameters:

– hydraulic loads on the barrier of between 0.03 and 3 m;

– hydraulic conductivity of between 10^{-10} and 10^{-8} m/s in the soil of the composite barrier (for PCC, GCC and ECC);

– soil thickness of between 0.3 and 5 m (for PCC, GCC and ECC);

– CGL thickness of between 6×10^{-3} and 14×10^{-3} m;

– hydraulic conductivity of the CGL between 10^{-12} and 10^{-10} m/s for virgin CGLs and a hydraulic conductivity greater than 10^{-10} m/s for CGLs aged through a cation exchange process, which may also be combined with hydration-dehydration cycles.

4.7. Points to remember

Clay geomembranes and geosynthetics are materials that carry out a barrier function and fulfill only this function.

They must be combined with other geosynthetic material or natural materials that can carry out the functions of drainage, filtration and especially protection, within a geosynthetic barrier system.

While a protective geotextile that also forms the support layer may be inserted under the geomembrane in a hydraulic structure, this is proscribed for the bottom and internal slopes of landfills, unless specified otherwise by the designer. The priority for waste storage facilities is ensuring good contact between the geomembrane and the passive barrier.

In the specific case of non-hazardous waste landfills, a GCL may be used to reinforce the passive barrier if the latter does not satisfy the regulatory requirements, without reducing the thickness of the barrier.

At present, geomembranes have no equivalent.

In the case of drainage systems, it may be possible to find an equivalence. In this case, the drainage geocomposite must be designed by taking into account the reduction factors in order to guarantee equivalence over the long term.

It is difficult to avoid defects when installing the geomembrane. Therefore, it is essential to carry out a quality control of the installation and plan for a defect detection process after the installation.

Defect	Contact condition	Empirical equation
Circular defect 2 < Φ < 20 mm	PCC (Giroud 1997)	$Q = 1.15h_w^{0.9}a^{0.1}k_s^{0.74}[1 + 0.1(h_w/H_s)^{0.95}]$
	GCC (Giroud 1997)	$Q = 0.21h_w^{0.9}a^{0.1}k_s^{0.74}[1 + 0.1(h_w/H_s)^{0.95}]$
	ECC (Touze-Foltz and Giroud 2003)	$Q = 0.096h_w^{0.9}a^{0.1}k_s^{0.74}[1 + 0.1(h_w/H_s)^{0.95}]$
	Virgin GM-CGL (Touze-Foltz and Barroso 2006)	$Q = 2 \times 10^{-4}h_w^{0.87}a^{0.07}k_s^{0.64}[1 + 0.31(h_w/H_s)^{0.79}]$
	Aged GM-CGL (Bannour and Touze 2019)	$Q = 9.405 \times 10^{-8}h_w^{0.91}a^{0.27}k_s^{0.26}\left[1 + 0.34\left(\frac{h_w}{H_s}\right)^{0.68}\right]$
Circular defect 100 < Φ < 600 mm	PCC (Giroud and Touze-Foltz 2005)	$Q = 2.60h_w^{0.84}a^{0,18}k_s^{0.77}\left[1 - 0.1\left(\frac{h_w}{H_s}\right)^{0.027}\right]$
	GCC (Giroud and Touze-Foltz 2005)	$Q = 0.64h_w^{0.84}a^{0.18}k_s^{0.77}\left[1 - 0.1\left(\frac{h_w}{H_s}\right)^{0.027}\right]$
	ECC (Giroud and Touze-Foltz 2005)	$Q = 0.33h_w^{0.84}a^{0.18}k_s^{0.77}\left[1 - 0.1\left(\frac{h_w}{H_s}\right)^{0.027}\right]$
	Virgin GM-CGL (Touze-Foltz and Barroso 2006)	$Q = 0.116h_w^{0.54}a^{0.4}k_s^{0.82}\left[1 - 0.22\left(\frac{h_w}{H_s}\right)^{-0.35}\right]$
	Aged GM-CG (Bannour and Touze 2019)	$Q = 3.03 \times 10^{-3}h_w^{0.65}a^{0.86}k_s^{0.64}\left[1 + 0.01\left(\frac{h_w}{H_s}\right)^{0.56}\right]$

Q = flow in m^3/s; h_w = hydraulic load over the geomembrane m; a = area of the circular defect in m^2 ; k_s = equivalent hydraulic conductivity of the soil in m/s; H_s = soil thickness.

Table 4.4. *Empirical equations to predict the leakage rate in the case of circular defects*

Defect	Contact condition	Empirical equation
Longitudinal defect 2 mm < *b* < 20 mm	ECC	$Q_L = 0.42 h_w^{0.45} b^{0.004} k_s^{0.87} \left[1 + 0.52\left(\frac{h_w}{H_s}\right)^{0.59}\right]$
	GCC	$Q_L = 0.65 h_w^{0.45} b^{0.004} k_s^{0.87} \left[1 + 0.52\left(\frac{h_w}{H_s}\right)^{0.59}\right]$
	PCC	$Q_L = 1.64 h_w^{0.45} b^{0.004} k_s^{0.87} \left[1 + 0.52\left(\frac{h_w}{H_s}\right)^{0.59}\right]$
Damaged wrinkle 100 mm < *b* < 600 mm	ECC	$Q_L = 0.63 h_w^{0.45} b^{0.1} k_s^{0.87} \left[1 + 0.282\left(\frac{h_w}{H_s}\right)^{0.82}\right]$
	GCC	$Q_L = 0.89 h_w^{0.45} b^{0.1} k_s^{0.87} \left[1 + 0.282\left(\frac{h_w}{H_s}\right)^{0.82}\right]$
	PCC	$Q_L = 1.98 h_w^{0.45} b^{0.1} k_s^{0.87} \left[1 + 0.282\left(\frac{h_w}{H_s}\right)^{0.82}\right]$

Q = flow (m^3/s); Q_L = flow per meter length of the longitudinal defect (m^2/s); h_w = hydraulic load on the geomembrane (m); a = area of the circular defect (m^2); b = width of the longitudinal defect or the damaged wrinkle (m); k_s = hydraulic conductivity of the soil (m/s); H_s = soil thickness.

Table 4.5. *Empirical equations to predict the leakage rate in the case of longitudinal defects (Touze-Foltz and Giroud 2003)*

4.8. References

AFNOR (2013). NF P84-500, Géomembranes – Dictionnaire des termes relatifs aux géomembranes. Standard, AFNOR, Paris.

AFNOR (2015). NF EN ISO 10318-1, Géosynthétiques – Partie 1 : Termes et définitions, Terminologie. Standard, AFNOR, Paris.

AFNOR (2017). NF G38-067, Géosynthétiques, géotextiles et produits apparentés – Stabilisation d'une couche mince sur pente – Justification du dimensionnement et éléments de conception. Standard, AFNOR, Paris.

AFNOR (2019a). NF EN 14150, Géomembrane – Détermination de la perméabilité aux liquides. Standard, AFNOR, Paris.

AFNOR (2019b). NF P84-700. Géosynthétiques bentonitiques – Généralités – Définitions. Standard, AFNOR, Paris.

AFNOR (2019c). NF P84-705, Géosynthétiques bentonitiques – Détermination à l'oedoperméamètre des caractéristiques de gonflement, flux, perméabilité des géosynthétiques bentonitiques (GSB) – Essais de caractérisation et essai de performance. Standard, AFNOR, Paris.

AFNOR (2020). NF EN ISO 12958, Géotextiles et produits apparentés – Détermination de la capacité de débit dans leur plan. Standard, AFNOR, Paris.

ASTM International (2003). ASTM D7002 – Standard practice for leak location on exposed géomembranes using water puddle system. Standard, ASTM, West Conshohocken.

ASTM International (2016). ASTM D7007 – Standard practices for locating leaks in géomembranes covered with water or earthen materials. Standard, ASTM, West Conshohocken.

Bannour, H. and Touze, N. (2019). Empirical equations for calculating the rate of liquid flow through GMB-GCL composite liners for GCLs subjected to cation exchange and wet dry cycles. In *17th African Regional Conference on Soil Mechanics and Geotechnical Engineering*. CRC Press, Boca Raton and Taylor and Francis, Cape Town.

Barral, C. and Touze-Foltz, N. (2015). Flow rate measurement in undamaged multicomponent geosynthetic clay liners. *Geosynthetics International*, 19(6), 491–496.

Board, M. and Laine, D.L. (1995). Coralling liner nightmares. *MSW Management: The Journal for Municipal Solid Waste Professionals*, 5(2), 48–50.

Charpentier, C. and Jacquelin, T. (2019). Détection de fuites sur géomembranes : observations et résultats provenant de cas récents. In *12èmes Rencontres Géosynthétiques*. Comité français des géosynthétiques, Nancy.

Colucci, P. and Lavagnolo, M.C. (1995). Three years field experience in electrical control of synthetic landfill liners. In *Proceedings 5th International Landfill Symposium*. Springer, Cagliari.

Comité français des géosynthétiques (1995). Recommandation générales pour l'utilisation des géosynthétiques dans les installations de stockage de déchets. Guide, CFG, Rueil-Malmaison [Online]. Available at: http://www.cfg.asso.fr/publications/guides-de-recommandations.

Comité français des géosynthétiques (1998). Recommandation générales pour la réalisation d'étanchéité par géosynthétiques bentonitiques. Guide, CFG, Rueil-Malmaison [Online]. Available at: http://www.cfg.asso.fr/publications/guides-de-recommandations.

Comité français des géosynthétiques (2002). Présentation des méthodes de détection de défauts dans les dispositifs d'étanchéité par géomembranes. Guide, CFG, Rueil-Malmaison [Online]. Available at: http://www.cfg.asso.fr/publications/guides-de-recommandations.

Comité français des géosynthétiques (2011). Recommandation pour l'utilisation des géosynthétiques bentonitiques en installations de stockage de déchets. Guide, CFG, Rueil-Malmaison [Online]. Available at: http://www.cfg.asso.fr/publications/guides-de-recommandations.

Comité français des géosynthétiques (2014). Recommandation pour l'emploi des géosynthétiques dans les systèmes de drainage et de filtration. Guide, CFG, Rueil-Malmaison [Online]. Available at: http://www.cfg.asso.fr/publications/guides-de-recommandations.

Comité français des géosynthétiques (2017). Recommandation générales pour la réalisation d'étanchéité par géomembranes. Guide, CFG, Rueil-Malmaison [Online]. Available at: http://www.cfg.asso.fr/publications/guides-de-recommandations.

Darilek, G.T. and Miller, L.V. (1998). Comparison of dye testing and electrical leak location testing of a solid waste liner system. In *6th International Conference on Geosynthetics*. Leak Location Services Inc. (LLSI), Atlanta.

Giroud, J.P. (1997). Equations for calculating the rate of liquid migration through composite liners due to geomembrane defects. *Geosynthetics International*, 4(3/4), 335–348.

Giroud, J.P. and Bonaparte, R. (1989). Leakage through liners constructed with geomembranes – Part II. Composite liners. *Geotextiles and Geomembranes*, 8(2), 71–111.

Giroud, J.P. and Plusquellec, H. (2017). Water, canals and geosynthetics. In *Proceedings of Geoafrica 2017*, Marrakech.

Giroud, J.P. and Touze-Foltz, N. (2005). Equations for calculating the rate of liquid flow through geomembrane defects of uniform width and finite or infinite length. *Geosynthetics International*, 12(4), 191–204.

Hsuan, G., Schroeder, H.F., Rowe, R.K., Müller, W., Greenwood, J., Cazzuffi, D., Koerner, R.M. (2008). Long-term performance and lifetime prediction of geosynthetics. In *Eurogeo 4*, Edinburgh.

International Geosynthetics Society (2009). Recommended descritpions of geosynthetics functions, geosynthteics terminology, mathematical and graphical symbols [Online]. Available at: https://www.geosyntheticssociety.org/wp-content/plugins/resources/documents.

Journal officiel de la république française (2016). Arrêté du 15 février 2016 relatif aux installations de stockage de déchets non dangereux. *Journal officiel de la république française (JORF)*, 0069, DEVP1519168A.

Koerner, R.M. (1999). *Designing with Geosynthetics*, 4th edition. Prentice Hall, Hoboken.

Laine, D.L. and Mosley, N.G. (1993). Leak location survey of a soil covered geomembrane at a landfill site in the UK. In *Green'93 Conference Proceedings*, Bolton.

Ministère de L'Écologie, de l'Énergie, du Développement durable et de l'Aménagement du territoire (2009). Guide de recommandations pour la conception et l'évaluation de dispositifs d'équivalence en étanchéité passive d'installation de stockage de déchets, version 2. Report, BRGM, Orléans.

Nosko, V. and Touze-Foltz, N. (2000). Geomembrane liner failure: Modelling of its influence on contaminant transfer. In *Eurogeo 2*, Bologna.

Rowe, R.K. (1998). Geosynthetics and the minimization of contaminant migration through barrier systems beneath solid waste. In *6th International Conference on Geosynthetics*, Atlanta.

Snow, M., Bishop, I., Keenan, R. (1999). Case history of geomembrane damage assessement. In *Geosynthetics '99*, Boston.

Stoltz, G., Croissant, D., Hérault, A. (2017). Étude sur le long terme du phénomène d'intrusion des filtres dans les géocomposites de drainage. In *Rencontres géosynthétiques 2017*, Lille.

Touze-Foltz, N. (2001). Modélisation des transferts advectifs dans les étanchéités composites de centres de stockage de déchets. PhD Thesis, École Nationale Supérieure des Mines de Paris, Paris.

Touze-Foltz, N. (2010). State of the art and durability insights regarding the use of geosynthetics for lining in hydraulic and environmental applications. In *9th International Conference on Geosynthetics*, Guarujá.

Touze-Foltz, N. (2018). Healing the world: A geosynthetics solution. In *11th International Conference on Geosynthetics*, Seoul.

Touze-Foltz, N. and Barroso, M. (2006). Empirical equations for calculating the rate of liquid flow through GCL-geomembrane composite liners. *Geosynthetics International*, 13(2), 73–82.

Touze-Foltz, N. and Giroud, J.P. (2003). Empirical equations for calculating the rate of liquid flow through composite liners due to geomembrane defects. *Geosynthetics International*, 10(6), 215–233.

Touze-Foltz, N., Barroso, M., Cartaud, F. (2006). Investigation des débits de fuite dans les étanchéités composites à l'échelle métrique. In *Rencontres Géosynthétiques 2006*, Montpellier.

Touze-Foltz, N., Mendes, M., Barral, C., Mazéas, L. (2015). Diffusive transport of phenolic compounds through two coextruded geomembranes. *International Journal of Geosynthetics and Ground Engineering*, 1(25). doi: 10.1007/s40891-015-0028-0.

Touze-Foltz, N., Bannour, H., Barral, C., Stoltz, G. (2016). A review of the performance of geosynthetics for environmental protection. *Geotextiles and Géomembranes*, 44(5), 656–672.

Touze-Foltz, N., Lupo, J., Barroso, M. (2018). Geoenvironmental applications of geosynthetics. In *Eurogeo 4*, Edinburgh.

Von Maubeuge, K.P., Sreenivas, K., Pohlmann, H. (2011). The new generation of geosynthetic clay liners. In *Geosynthetics India' 11 and an Introductory Course on Geosynthetics*, Chennai.

White, C. and Barker, R.D. (1997). Electrical leak detection system for landfill liners: A case history. *Groundwater Monitoring & Remediation (GWMR)*, 17(3), 153–159.

Zanzinger, H. and Touze-Foltz, N. (2009). Performance des géosynthétiques bentonitiques en couvertures d'installations de stockage de déchets. In *Rencontres Géosynthtéiques 2009*, Nantes.

5

Geosynthetic Filters: Uses, Feedback from Experience and Current Research

Guillaume STOLTZ
RECOVER, INRAE, Aix-Marseille Université, Aix-en-Provence, France

5.1. Introduction

Filtration is one of the primary functions for which geosynthetics are used in soil and, more generally, in any structure. As Touze-Foltz (2016) highlights, this usage dates back to the winter storms of 1952–1953 in the Netherlands, when woven textiles were used to replace granular filters. In April 1977, France organized the first "International Conference on the use of Textiles in Geotechnology", at a time when research into this was still new and the terms "geotextile" and "geomembrane" had not yet been formalized. Of the eight sessions in this conference, one was dedicated specifically to "The functioning of filters containing textiles". This first conference was succeeded by other international conferences on geosynthetics, with smaller regional and even international conferences specifically on "Geofilters", the term used at the time. While still in use today, this is not the term used in the international standard for terms and definitions in the field of geosynthetics BS EN ISO 10318-1 (BSI 2015a) nor is it used in the terminology recommended by the International Geosynthetics Society (IGS) (2009). The four international *Geofilter* conferences were held in Karlsruhe (Germany) in 1992, in Montreal (Quebec) in 1996, in Warsaw (Poland) in 2000 and finally in Stellenbosch (South Africa) in 2004. Several of the references in this chapter are from these conferences. Geosynthetic filters today are largely used to replace traditional granular filters, or even in association with these. In addition to their proven efficacy, there are many advantages to using geosynthetic filters. These include quality control of the product in the factory or in an independent laboratory, the reduced sampling of natural materials, a reduction in energy and environmental costs linked to the transport of materials and the installation on the site through certified companies (quality control and quality assurance on work sites).

This chapter, dedicated to geosynthetic filters, does not aim to carry out an exhaustive comparison between granular and geosynthetic filters, nor does it aim to present all the research carried out on geosynthetic filters, nor even to list out all the design methods used for geosynthetic filters. There are articles available in the reference list at the end of this chapter that summarize the state-of-the art knowledge on geosynthetic filters. For instance, Giroud (1996) offers an exhaustive comparison between granular filters and geosynthetic filters, and Moraci (2010), Aydilek (2012) and Cazzuffi et al. (2016) present an exhaustive analysis of the design methods used for geosynthetic filters.

The objective of the present chapter is to give an overview of geosynthetic filters, the conventionally used design principles, the causes for failure over time, examples of feedback from experience after several years in service and, finally, some areas for future research.

5.2. General information on filtration using geosynthetics

5.2.1. *Filtration function*

5.2.1.1. *Definitions*

In scientific bibliography, Giroud (1984) defined the function of a filter by clearly differentiating two points of view:

– A geosynthetic filter, placed across a flow of liquid with fine suspended particles, makes it possible to retain most fine particles, while allowing the liquid to flow through (1).

– A geosynthetic filter, placed in contact with the soil, allows water to percolate through the soil, while blocking the passage of most soil particles entrained by the water's hydrodynamic forces (2).

It is interesting to note that Giroud (1981) had added the concept of a filter associated with an armor. In this case, the filter is placed between the soil that must be protected from wave impact and a protective layer (rocks, concrete blocks, gabions, etc.). Unlike the two filtration applications, (1) and (2), here the flow is non-continuous and dynamic, and the directions may switch from one to the other, and the filter confinement may be locally heterogeneous.

In definitions (1) and (2), the geosynthetic carries out the filtration function perpendicular to its plane. When considering only the retention of solid particles, the filtration function may be called a mechanical function, although it makes it possible to control the flow of a fluid, whose properties are impacted (loss of hydraulic head; reduced viscosity, in the case of a sludge made up of water and fine clay traveling through the filter, for example).

Today, the filtration function of geosynthetics is defined in the international standard BS EN ISO 10318-1 (BSI 2015a) as follows: restraining of uncontrolled passage of soil or other particles subjected to hydrodynamic forces, while allowing the passage of fluids into or across a geosynthetic material. This definition covers both definitions given in Giroud (1984) but loses the nuance around the object being filtered: a fluid or a soil, which appear, respectively, in definitions (1) and (2) in Giroud (1984). And yet, this is a major difference depending on the purpose of the geosynthetic filter. The use of the filter in definition (1) in Giroud (1984) is similar to the most common use of filters, used in industry and many day-to-day devices. Their objective is to separate all solid particles from the fluid and consequently, the filter must be replaced (or cleaned) when the "cake" of particles formed upstream or within filter no longer allows satisfactory fluid flow. This type of filtration will be seen further on in this chapter. On the other hand, the use of a filter as in definition (2) in Giroud (1984) is unusual, as it implies a very long-term functioning of the filter. Indeed, the use of geosynthetic filters in civil engineering projects does assume they have a long-lasting behavior which, in other words, would exclude any regular replacement of the filter when it stops functioning in a satisfactory manner. This will be explained in section 5.3.1, where we look at the mechanisms that allow the geosynthetic filter to behave in this way when used in soil.

5.2.1.2. *Pictogram*

The standard BS EN ISO 10318-2 (BSI 2015b) defines a pictogram representing the filtration function (Figure 5.1). The arrows represent a mechanical stress from the in situ soil, on one side, and from another soil or a synthetic drainage core on the other side. This figure does not cover all use cases, such as the use of geosynthetic filters in the shape of a tube to dewater sludge composed of water and solid particles in suspension. Indeed, in this specific case, only hydraulic/mechanical pressure is applied to one side of the geosynthetic filter.

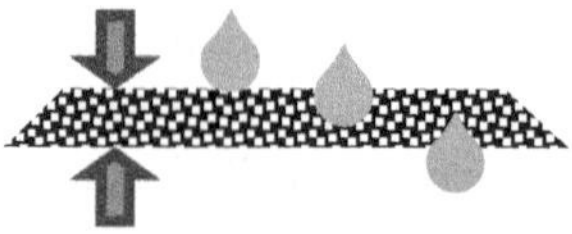

Figure 5.1. *Pictogram of the filtration function as per the standard BS EN ISO 10318-2 (BSI 2015b). For a color version of this figure, see www.iste.co.uk/villard/geosynthetics.zip*

5.2.2. *Filter geosynthetics and drainage geocomposites*

5.2.2.1. *Geotextile filters*

Filter geosynthetics can also be called geosynthetic filters or geotextile filters, like the "paper filters" that are traditionally used in laboratories. You may also come across the term "filtration geosynthetics" or "filtration geotextiles". Geosynthetics that can carry out filtration are obviously permeable to water. The products that can perform this function are generally geotextiles, and a few examples of these are given in Figure 5.2.

Most geotextile filters installed in structures today are directly connected to a synthetic drainage core. These composite products make it possible to perform two functions and will be defined in section 5.2.2.3. In some cases, other elements, such as reinforcement fibers, may also be combined with geotextile filters, so as to perform the double function of filtration and reinforcement.

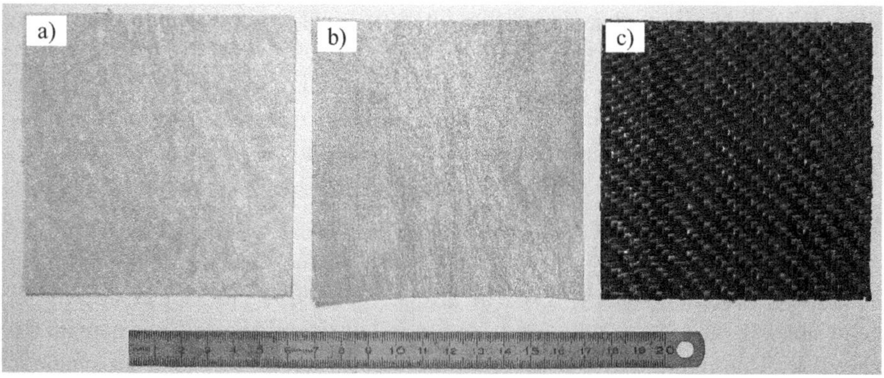

Figure 5.2. *Examples of geotextile filters: (a) needle-punched non-woven; (b) thermal-bonded non-woven; (c) woven strips (source: INRAE). For a color version of this figure, see www.iste.co.uk/villard/geosynthetics.zip*

Finally, geocells are a unique case as these products can fulfill stabilization functions or prevent surface erosion, while also filtering the compacted soil within. Considering the rapid innovation in this domain, it is difficult to provide an exhaustive list of all composite, multi-functional products that can perform filtration. By making use of the main concepts of soil filtration using geosynthetics, designers and users will be able to identify this function in a composite geosynthetic and will be able to apply design rules. In the rest of this chapter, the term "geotextile filter" is chosen over the more generic "geosynthetic filter". This is further justified when you consider that the element that carries out the filtration is ultimately the geotextile component and that the other functions cited earlier are secondary (at least from the point of view of this chapter).

5.2.2.2. *Structure of geotextile filters*

In general, geotextile filters are manufactured from polypropylene and/or polyester fibers. There are also products made from natural fibers (e.g. hemp and linen), which are much older than their synthetic-fiber counterparts. These textiles may be woven, non-woven or knitted.

Woven geotextiles are generally designed from polypropylene and/or polyester strips or even strips of natural materials, which may be monofilament or multifilament, and of the same size or variable sizes (Figure 5.3). The filtration opening size (section 5.3.2.1) can vary between 50 μm and 2 mm.

Non-woven geotextiles may be manufactured from continuous or short fibers, which are generally synthetic (polypropylene or polyester), and assembled in the form of sheets through needle-punching or thermal bonding (Figure 5.4). Their filtration opening size can go up to 500 μm.

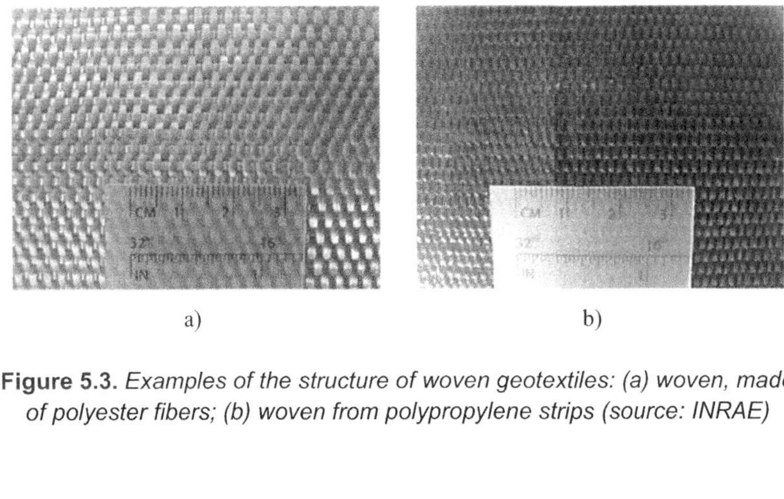

a) b)

Figure 5.3. *Examples of the structure of woven geotextiles: (a) woven, made of polyester fibers; (b) woven from polypropylene strips (source: INRAE)*

Woven geotextiles are manufactured from synthetic fibers (e.g. polyester) or natural fibers and have very variable filtration opening sizes (Figure 5.5).

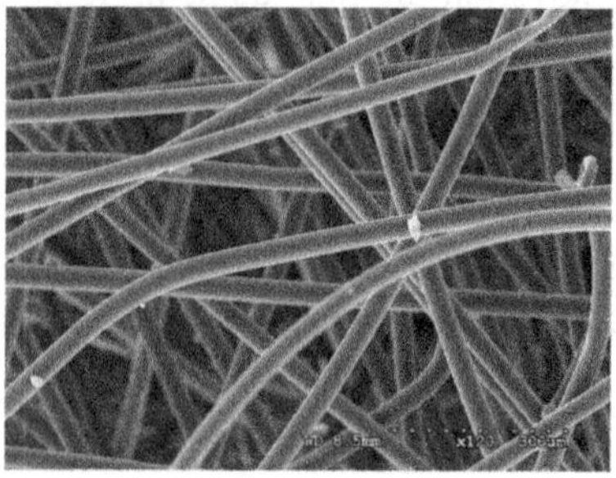

Figure 5.4. *Example of the structure of a non-woven geotextile with fibers assembled through needle-punching (source: INRAE)*

Figure 5.5. *Example of the structure of a knitted textile (source: INRAE)*

5.2.2.3. *Drainage geocomposites*

Although this chapter is dedicated to geotextile filters, given that these are mainly used in soil drainage applications, it would be helpful to briefly introduce drainage geocomposites. There is a wide variety of drainage cores and they can additionally be combined in multiple ways with one or more geotextiles: glueing, heat-welding, stitching, etc. (Figure 5.6). These composite products, made up of at least one geotextile filter and one drainage core, are commonly called drainage geocomposites or geocomposite drains. The term "geocomposite" is defined in the standard BS EN ISO 10318-1 (BSI 2015a) and the term "drainage geocomposite" is defined in the terminology recommended by the International Geosynthetics Society (IGS) (International Geosynthetics Society 2009). Thus, a drainage geocomposite is a geosynthetic that performs both filtration and drainage. The drainage function is defined in the standard BS EN ISO 10318-1 (BSI 2015a) as "collecting and transporting of precipitation, ground water, and/or other fluids in the plane of a geosynthetic material". The design of a drainage geocomposite therefore involves a double design: one for the geotextile filter and another for the drainage part. With regard to the drainage function, the reader can refer to the design standard NF G 38-061 "Use of geotextiles and geotextiles related products – Drainage and filtration systems – Justification of design and design elements" (AFNOR 2017) or the recommendation guide published by the *Comité français des géosynthétiques* (CFG) regarding the use of geosynthetics in drainage and filtration systems (Comité français des géosynthétiques 2014).

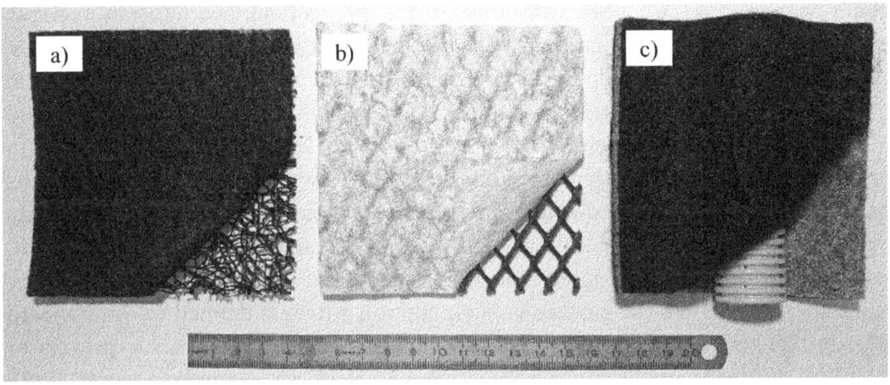

Figure 5.6. *Examples of drainage geocomposites composed by (a) and (b) a geotextile filter and a uniform drainage core or (c) a geotextile filter and drainage core plus mini-drains (source: INRAE). For a color version of this figure, see www.iste.co.uk/villard/geosynthetics.zip*

5.2.3. *The role of geotextile filters and their applications*

5.2.3.1. *The role of geotextile filters*

As defined earlier, the role of geotextile filters in the soil is to carry out filtration by retaining solid particles and allowing the passage of water. This filtration function can have different objectives:

– When combined with a drainage system (granular system and/or with a synthetic drainage core) that aims to reduce interstitial pressures in the soil, the geotextile filter can help prevent the drainage core being contaminated by the fine soil particles formed from eroded particles and entrained by hydrodynamic forces. In this configuration, the objective of the drainage system is to reduce interstitial pressures in the soil and therefore the geotextile filter must not induce any loss of water load as the water passes through.

– When placed between two compacted soils of different natures, the geotextile filter can prevent fine, eroded particles from being transported by the water (suffusion) from the upstream soil to the downstream soil. This prevents destabilization of the soil. In this configuration, the geotextile filter performs both separation and filtration functions. In addition to preventing a possible erosion of the upstream soil through suffusion, the geotextile filter also makes it possible to prevent erosion through piping when the upstream soil might offer preferential flow paths.

– When placed between a soil and rip-rap or gabions (when protecting banks along waterways), the geotextile filter helps prevent surface erosion (contact erosion) of the soil particles that make up the structure. As specified in Giroud (2010), when used for this application, the geotextile must not get clogged. This is to prevent instabilities in the soil being protected, especially in the case of a sudden dip in water levels. It is therefore acceptable for particles retained by the filter to undergo lessivage through dynamic and alternating flows, which results in a gradual erosion of the soil. Consequently, in this application, the geotextile filter only partially retains particles, and maintenance operations must be carried out on these structures. It must also be noted that in this case there is a difference between the filtration function and the surface erosion control, which is defined in the standard BS EN ISO 10318-1 (BSI 2015a) as: "use of a geosynthetic materials to prevent or limit soil or other particle movements at the surface of, for example, a slope". Both these functions are quite close and have the same objective of stabilizing soil, that is, preventing the movement of particles due to hydrodynamic forces across or in the plane of the filter, which is permeable to water. The concept of surface erosion control is primarily used in the context of hydrodynamic forces arising from precipitation or runoff.

– Used in the form of tubes, the geotextile filter makes it possible to separate water from the solid fraction of sludge made up of water and fine, suspended particles. This process is conventionally called drying or dewatering of sludge using geotextiles tubes.

– Used in the form of a retention or filtration barrier, the geotextile filter prevents water carrying fine particles from flowing to the edges of construction sites.

In the first three points above, the geotextile filter helps stabilize the upstream soil and the filtration function thus corresponds to definition (1) in Giroud (1984) (section 5.2.1.1), while in the last two points, the geotextile filter's chief function is to hold back the majority of fine, suspended particles in a flow, and the filtration function then corresponds to definition (2) in Giroud (1984) (section 5.2.1.1).

5.2.3.2. *Applications*

In the case of the filtration of compacted soils, geotextile filters can be used in many types of structures, such as hydraulic structures (dams, basins, canals, dikes etc.), earthwork (excavation or filling), transport (roadways, railways, airport runways, etc.), engineering structures (bridges, retaining structures, tunnels, etc.) and their foundation soil (accelerating consolidation in fine soils), foundations for buildings, and hazardous and non-hazardous waste storage facilities. The CFG recommendations guide (Comité français des géosynthétiques 2014) gives many examples and illustrations of how geotextile filters are used in different types of structures.

The filtration of water carrying fine particles (sludge) has already been mentioned previously. This filtration is generally carried out for temporary applications, that is, to filter sediment-laden water at a work-site, dewater sediments from dredging, etc. Unlike the filtration of in situ soils, where the quantity of water to be filtered may be considered to be unlimited, the filtration of sludge implies that there is a limited quantity of water to be filtered.

5.2.4. *Reference documents for design*

There are many documents, recommendation guides and standards today that allow end-users to choose the geotextile filter that is best-suited to the structure being considered. These documents have details on the various materials used, the stresses to which the geotextile filter will be subject, and, finally, the critical level as well as the in-service period of the structure. The *Comité français des géosynthétiques* (CFG) was one of the first learned societies to lay down recommendations on the use of geotextile filters, in a booklet published in April 1986. In 2014, this booklet was revised (Comité français des géosynthétiques 2014) and has been made freely available to all on the CFG website. This guide also refers readers to other recommendations and standards for very specific applications. In February 1993, the French Geosynthetic Standardization Commission published the standard G 38-061, specifying the "Recommendations for the use of geotextiles and geotextile-related products – Determining hydraulic characteristics and installing geotextiles and geotextile-related products used in drainage and filtration systems". This standard was significantly revised in 2014 and became the standard NF G 38-061 (AFNOR 2017). Finally, the International Standardization Organization (ISO) appointed an international group of experts to draw up the standard 18228-3 (ISO 2021) related to recommendations on the use and design of geosynthetic filters. All the documents cited here have been drawn up by international groups of experts. This present chapter does not aim to present an exhaustive review of all these documents, nor to replace them as a new reference document. Instead, it aims to illustrate the general design principles and put them in perspective in relation to current and future research.

5.3. Design methodology (compacted soils)

5.3.1. *General principles*

The general principles described here are given for the case of the filtration of compacted soils (or in situ soils) and not for cases requiring the filtration of fine suspended particles in a flow (which is presented in section 5.4). To filter in situ soils, the geotextile filter must be in close contact with the soil. This is an essential condition that must be guaranteed from the time of installation and throughout the in-service period of the geotextile filter.

The performance of a geotextile filter in soil is determined by two properties, which may appear to be in conflict from some viewpoints:

– The geotextile filter must be sufficiently closed (in terms of pore size) to retain soil particles that form the soil skeleton and prevent lessivage through internal erosion and/or to prevent the pollution of the drainage system downstream of the filter.

– On the other hand, the geotextile filter must be open enough to allow the passage of the finest unstable particles, in order to prevent clogging, and allow water to pass perpendicular to its plane without any significant loss in water load, in order to avoid an increase in the interstitial pressure in the downstream soils.

As stated in section 5.2.1.1, filtration using geotextiles in geotechnical structures and civil engineering structures requires the filtration function to have a durability that is at least the same as that of the structure. This durability is ensured by the formation of a stable, self-filtration granular layer in the soil, made up of the coarsest elements (representative of the granular skeleton) and maintained by geometric blocking between the fibers, or through an arching effect (bridging). As shown in Figure 5.7, after the lessivage of unstable fine particles, the stable bridges are formed by the stable coarse particles, which will themselves retain the fine upstream particles. After a transition period, the geotextile filter will enable the soil to carry out self-filtration by creating a stable transition zone in the vicinity of the geotextile, with transition granulometry and a permeability that is greater than that of the base soil. After the geotextile filter is installed in the structure, the first part of the filtration is a transition period when the fine particles in the filtration region are entrained by the flow, which leads to water in the drainage system being slightly laden with fine particles. After this transition phase, only clear water flows out of the filtration system.

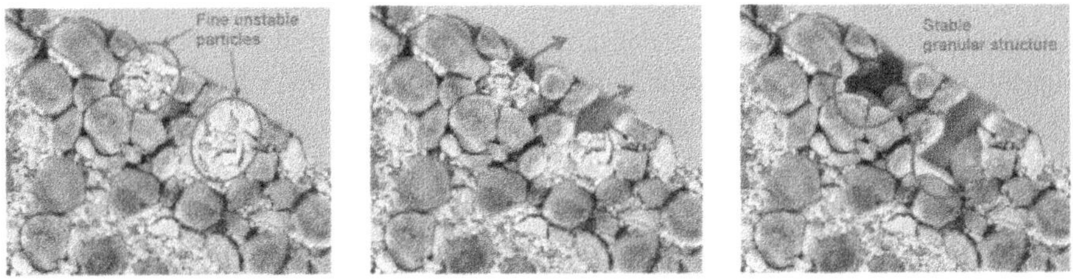

Figure 5.7. *The mechanism for the self-filtration of soil, enabled by the geotextile filter after the removal of the unstable, fine particles: evolution of the soil–geotextile interface (source: Comité français des géosynthétiques 2014). For a color version of this figure, see www.iste.co.uk/villard/geosynthetics.zip*

When selecting a geotextile filter, many parameters must be taken into account. These are related to:

– the site conditions, for example, the water flow (laminar or turbulent, unidirectional or bidirectional, permanent or cyclic, etc.), the water turbidity, the physicochemical conditions of the ground (section 5.3.4) and the mechanical actions (static or dynamic mechanical stresses in the structure, as in the case of geotextile filters under rip-rap, used to protect banks; the presence of roots; etc.);

– the installation conditions that could cause damage through mechanical or physicochemical conditions (oxidation through ultraviolet (UV) radiation, etc.).

In addition to these factors, it is essential to have precise knowledge on the soil to be filtered. This includes its granulometry (and variations in the granulometry), the relative density, the permeability of the soil, as well as the cohesion and dispersibility of its fine fraction.

It is also important to verify how stable the soil is against suffusion. This can be done using the criterion from (Kenney and Lau 1985), for instance, which is based on a graphic method that uses the granulometric curve of the soil. Cazzuffi et al. (2016) have presented a detailed analysis of the internal stability of the soil, as well as a recent method to verify the compatibility of a geotextile filter used to filter a soil with an internally unstable soil skeleton.

5.3.2. *Essential characteristics of geotextile filters*

5.3.2.1. *Filtration opening size*

The opening size of a geotextile filter is the dimension of the largest opening, that is, the dimension of the largest particle that can pass through the product. As it is not realistic or feasible to determine the dimension of the largest opening, a slightly smaller dimension is usually considered. Giroud (1981) defined the opening size, O_{95}, based on the size distribution of the opening sizes of the geotextile filter. Thus, O_{95} refers to a value where 95% of the pores are smaller than or equal to this dimension. The theoretical approach used by Fayoux and Evon (1982), based on a hypothesis where the fibers have a periodic distribution, made it possible to assess an order of magnitude for this parameter. However, based on tests that were developed to determine opening sizes for geotextile filters, the concept used today is not the same as the concept proposed by Giroud (1981). For instance, it may involve the dry-sieving of glass beads, as in the standard ASTM D4751 (ASTM International 2021), or the wet-sieving of standard sand, as in the standard BS EN ISO 12956 (BSI 2020). In ASTM D4751, O_{95} is evaluated from the percentage of a mass of glass beads lying within one interval. In other terms, O_{95} is the dimension for which 95% of the particles (for a given internal) are retained by the geotextile filter (5% of the particles pass through). Despite using a similar notation, this definition is different from the definition used in Giroud (1981), which is based on the size distribution of the openings. The terminology recommended by the International Geosynthetics Society (IGS) (International Geosynthetics Society 2009) goes back to this concept and defines the apparent opening size as the particle size of a standard sand fraction for which a specified percentage by mass would be retained, as determined by the dry-sieving method. In the standard BS EN ISO 12956 (BSI 2020), O_{90} is assessed based on the d_{90} of the standardized fraction of sand that passes through the geotextile filter. Thus, in the standard BS EN ISO 10318-1 (BSI 2015a), which defines terminology related to geosynthetics, the opening size is defined as the size of opening that corresponds to the maximum particle size of 90% by weight of the soil passing through the geotextile. Ultimately, although used for the same concept of filtration opening size, these different definitions and the different evaluation methods associated with them lead to different values for the same geotextile filter.

In the study carried out by Dierickx and Myles (1996), several methods were compared, especially the dry-sieving and wet-sieving methods, which were carried out in different laboratories. This inter-laboratory campaign showed that the wet-sieving method, described in the standard BS EN ISO 12956 (BSI 2020), was more reproducible, and determining the O_{90} value gave more consistent results than for O_{95}.

In practical terms, the filtration opening of the geotextile is necessarily bounded. Indeed, an opening size of several millimeters serves no purpose from a retention point of view, and conversely, a micrometric opening size would cause too much resistance to water flow to allow the filtration of in situ soil or the filtration of water laden with fine particles. The standard BS EN ISO 12956 (BSI 2020) specifies that the measured value of the opening size must be between the d_{20} and d_{80} obtained from the initial granulometry (before passing through the test filter) of the sand used for the measurement. Based on the granulometric curve defined in the standard, the minimum and maximum values are thus bounded between 50 and 600 μm. There are geotextile filters, especially thermal-bonded non-woven products, whose opening size is smaller than 50 μm, and these values cannot be accurately measured as per the standard BS EN ISO 12956 (BSI 2020). Therefore, performance tests must be carried out when using this kind of product.

5.3.2.2. *Flow velocity*

Initially, a geotextile filter's capacity for allowing water to pass through it was measured using a coefficient for permeability normal to the plane of the product (unit: $m.s^{-1}$) taken from Darcy's law, or more precisely, Darcy's equation. This coefficient of permeability corresponds to the ratio between the velocity and the hydraulic gradient, for a uniform and laminar flow at the standard temperature for the fluid. Permittivity, denoted by ψ (unit s^{-1}), was also defined from this coefficient, in the standard, BS EN ISO 10318-1 (BSI 2015a), corresponding to the coefficient of permeability divided by the thickness of the geotextile filter. In other words, permittivity is the volumetric flow rate of water, per unit area per unit head loss, under laminar flow conditions, normal to the plane of a product. However, for many geotextile products, the hydraulic regime is not laminar in the range of gradients that occur in soil, that is, Darcy's equation is not satisfied. Consequently, today we use the concept of flow velocity for a geotextile filter, allowing us to characterize the head loss caused by the product as water flows across it. The standard BS EN ISO 11058 (BSI 2019) gives a standardized protocol to measure this flow velocity, called the velocity index, fixing the head loss at 50 mm.

The standard BS EN ISO 10776 (BSI 2012a) offers a methodology to measure the flow velocity in the case where a normal stress is applied on the product. As with the case for a filtration opening under stress, at present there is no consensus on how to use this data for design purposes.

5.3.2.3. *Resistance to water penetration*

Resistance to water penetration is the minimum hydraulic head required to set up a water flow. This can be measured as per the standard BS EN 13562 (BSI 2000).

Although this characteristic is used in recommendation guides (Comité français des géosynthétiques 2014) as a design criterion, in practice it is very rarely measured for geotextile filters. This can be explained by the fact that this resistance to water penetration is partly due to the oils and lubricants used during the manufacturing process of these products. Once in contact with soils, which are generally damp, and as water passes through the filter, these oils are washed away and this reduces the resistance to water penetration, which is measured on the product fresh out of the factory. Consequently, this measurement could be not necessarily relevant over the long term. On the contrary, it is better for the user to verify with the manufacturer that no hydrophobic treatment has been used on the product's textile fibers.

5.3.2.4. *Other characteristics of geotextile filters*

There are other properties of geotextiles that are related to the filtration function, such as:

– The number of constrictions: to define this parameter, the reader can imagine that a soil particle, transported by the water flow across the thickness of a geotextile filtration system, travels through many successive "windows" delimited by fibers. Each of these windows is called a constriction. All along its way through the geotextile, the particle encounters a given number of constrictions. There used to be a French standard (XP G 38-030) to determine the number of constrictions. But while this characteristic makes sense for non-woven needle-punched geotextiles where the number of constrictions increase with thickness, it is not very meaningful for very fine, thermal-bonded geotextiles whose thickness can be smaller than 1 mm. The French standard is now cancelled and the number of constrictions is no longer used to dimension geotextile filters in compacted soils.

– Percent open area: as its name indicates, this parameter corresponds to the ratio between the open surface (openings) and the area covered by fibers (matter). These data are only useful for woven products. In France, there is no standardized method to carry out this measurement and these data are not used for design.

5.3.3. *Design criteria*

5.3.3.1. *Soil skeleton retention criterion*

In the case of the filtration of in situ or compacted soils, the primary function of the geotextile filter is to retain the soil skeleton formed by the coarsest particles. This criterion aims to maintain "only" those soil particles that withstand any possible form of soil erosion (contact erosion, through suffusion, piping, etc.) in order to prevent instabilities in the soil and/or clogging in the drainage system, if there is one. We use "only" to highlight that the unstable, fine particles, which do not contribute to the mechanical stability of the soil, must be able to pass through the filter in order to prevent upstream clogging or internal clogging of the filter. In international bibliographies, this criterion is generally called *retention criteria*, but we also come across *piping criteria*. This criterion is generally expressed in the following form:

$$O_f \leq C_R \times D_i \quad [5.1]$$

with:

– O_f as the opening size of the filtration geotextile, which can be determined using the standard BS EN 12956 (BSI 2020), for example;

– C_R as a coefficient (of retention), which takes into account the nature of the soil to be filtered and its characteristics, the hydraulic gradient and the role of the geosynthetic (only filter, or filter-drain); depending on the case, this coefficient may be larger than or smaller than 1;

– D_i as one of the characteristic diameters of the soil to be filtered, which is fixed depending on the nature of the soil. D_i can also be D_c, the critical suffusion diameter for soils that are unstable vis-à-vis suffusion (Cazzuffi et al. 2016).

Many authors have proposed values for C_R and D_i. Palmeira and Fannin (2002), and more recently Aydilek (2012), carried out a bibliographic synthesis of these studies. In France, the CFG recommendations (Comité français des géosynthétiques 2014) and the French standard governing the design of drainage/filtration geosynthetics (AFNOR 2017) provide design values for this criterion.

5.3.3.2. *Criterion on the non-retention of fine particles*

Although the geotextile filter must retain the particles that form the solid soil skeleton, it must also allow unstable, fine particles to pass through and in the long term this could lead to the clogging of the filtration system (sections 5.3.1 and 5.3.4.3), which comprises the soil–geotextile interface (or transition zone) and the geotextile filter itself. Clogging could lead to a significant increase in interstitial pressure in the soil, and thus provoke instabilities. In international scientific bibliographies, this criterion is generally called the *non-retention criteria of fines in suspension*, but we also encounter the term *clogging criteria*. This is expressed by the general form:

$$O_f \geq C_{NR} \times D_j \quad [5.2]$$

where:

– O_f is the opening size of the filtration geotextile, determined by the standard BS EN ISO 12956 (BSI 2020), for example;

– C_{NR} is a coefficient (of non-retention);

– D_j is one of the characteristic diameters of the soil to be filtered, which is fixed depending on the nature of the soil.

5.3.3.3. *Permeability criterion*

The permeability criterion makes it possible to impose a minimum resistance to water flow upon the geotextile filter. In other words, the presence of the filter must not generate head loss compared to a situation where a soil of the same nature as that to be filtered replaces the geotextile filter. Thus, extending this comparison of the same distance being covered by the water flow, through only the soil, in one situation, and through the soil and then the filter in another, for the same upstream head the presence of the filter must not lead to a reduced flow. The permeability criterion is expressed in the general form:

$$V_F \geq C_P \times k_s \times i_s \quad [5.3]$$

where:

– V_F is the flow velocity obtained through the geotextile for a head loss of 50 mm (which can be measured as per the standard BN EN ISO 11058 (BSI 2019), section 5.3.2.2);

– k_s is the hydraulic conductivity of the soil in m.s^{-1};

– C_P is a coefficient that mainly takes into account the critical level of the structure;

– i_s is the hydraulic gradient of the flow in the soil in the vicinity of the geotextile filter.

It is interesting to note that today, this criterion considers the flow velocity in the geotextile filter, while the early permeability criteria looked at the permittivity of the product. Giroud (2010) noted that the permeability criterion must also respect two conditions:

– An interstitial pressure condition: ideally, the presence of the geotextile filter should not lead to increased interstitial pressure in the soil. This condition is obtained if the permeability normal to the plane of the geotextile is greater than or equal to the hydraulic permeability of the soil multiplied by the hydraulic gradient in the soil.

– A flow rate condition: even when it is very permeable, the presence of a geotextile filter reduces the water flow as compared to a situation without a geotextile. Calculations using Darcy's equation showed that the decrease in flow rate does not exceed 10% (compared to the case without the geotextile) if the permeability of the geotextile filter is greater than or equal to the permeability of the soil (for products whose thickness is between 1 and 10 mm).

Of these two conditions, it seems that the most restrictive condition depends on the hydraulic gradient in the soil: for a gradient larger than 1, the most restrictive condition will be the "interstitial pressure" condition, and for a gradient smaller than 1, the most restrictive condition will be one related to flow rate. With respect to the conditions given in Giroud (2010), in equation [5.3], we consider the velocity index of the product measured for a head loss of 50 mm, which is the same as tolerating an increase of 50 mm in the soil interstitial pressure, in terms of the height of the water column, induced by the presence of the geotextile filter. Therefore, depending on the value of the hydraulic gradient in the soil (especially if it is smaller than 1) and depending on the coefficient C_p, the conditions in Giroud (2010) may be more restrictive, even if Giroud (2010) does not include the criterion of a safety coefficient.

5.3.3.4. *Other criteria*

We may find other criteria in recommendation guides or in the studies given in the bibliography. For example:

– The water penetration criterion: this criterion, given in the CFG recommendations (Comité français des géosynthétiques 2014) and in the French standard NF G 38-061 on the design of drainage/filtration geosynthetics (AFNOR 2017), uses the value of the resistance to water (section 5.3.2.3). Resistance to water penetration is related to the wettability of the geotextile, which depends on the nature of the surface (geotextile filter) as well as the nature of the fluid (generally water but may contain agents that affect surface tension and viscosity). In the two references given above, this value must be smaller than 5 m in terms of hydraulic head. However, this value was fixed quite arbitrarily by the experts who drew up the recommendations, as there is no scientific study in the bibliography that exhaustively shows the impact of resistance to water penetration on the performance of the filtration function in real usage conditions, whether in the short term or the long term. Indeed, as specified in section 5.3.2.3, the resistance to water penetration measured on a product fresh from the factory may largely be due to the manufacturing oils, which are hydrophobic, and which are gradually washed away over time. Consequently, the criterion of resistance to water penetration is given for safety purposes. Geotextile filters that do not satisfy this criterion can be used based on performance tests.

– The soil–geotextile filter contact criterion: the condition for close contact between the soil to be filtered and the geotextile filter is essential with respect to the durability of the filtration system. This condition notably depends on the density of the soil to be filtered, on the stiffness of the geotextile and is highly dependent on the installation of the product. Cazzuffi et al. (2002) have shown that if close contact between the soil and geotextile filter was not maintained, then in conditions of non-permanent and turbulent flow, a geotextile filter could not perform its function. Giroud (2005) theoretically demonstrated that a fine layer of fine particles could result in a drastic drop in the permeability of the filtration system. Although this criterion is essential for the durability of the system, at present there are no numbers for this criterion, which opens up a future field of research, for example, to define a method to inspect the installation, define a minimum confining stress, etc.

– Giroud (2010) proposed that the design of geotextile filters was based on three or four criteria, depending on the nature of the geotextile filter. In addition to the criteria on retention and permeability, common to all geotextile filters, Giroud (2010) suggested adding a porosity criterion and thickness criterion for non-woven, needle-punched geotextiles. The reason for these two additional criteria is that machines can manufacture products that may have very few openings, or be very fine, while still presenting opening sizes and a velocity index that fulfill the design criteria for retention and permeability. Therefore, the porosity criterion defined by Giroud (2010) makes it possible to guarantee that the geotextile filter has a sufficient number of openings, while the thickness criteria makes it possible to ensure that the geotextile filter is thick enough so that a small variation in thickness on the surface of the product does not lead to significant disparities in terms of filtration openings. Giroud (2010) suggested choosing a geotextile filter whose openings size was closest to the opening that respected the retention criterion and that this would implicitly ensure the fulfillment of the criterion of non-retention of unstable fine particles.

5.3.4. *Causes of failure*

5.3.4.1. *Improper installation*

While improper installation may be an obvious cause of failure, it is still an important one. All recommendation guides and/or design standards mention best practices for installing geotextile filters: inspecting the products when they are received on the work site, proper storage conditions before deploying the product, appropriate climatic conditions for installation by qualified personnel, sufficient covering of the layers, ensuring that the geotextile and soil to be filtered are in close contact, reduced exposure to UV radiation (respecting the information regarding durability given in the declarations of performance of the product), etc. It must be recalled that the smallest runoff of water laden with fine particles, any mechanical damage during installation, or the absence of any covering, even in localized zones, may have a long-term effect on the functioning of the geotextile filter and any drainage system that may be associated with it. For example, Palmeira and Galvis (2017) quantified the partial clogging of geotextile filters by defining an impregnation factor, which corresponds to the mass of solid particles in the pores of the geotextile divided by the mass of the geotextile filter. Through sampling from the site, they showed that this factor could have values going up to 10.

5.3.4.2. *Physicochemical alteration*

Depending on the nature of their constituent polymer, the fibers of a geotextile may be altered by physicochemical conditions such as the presence of UV radiation, effects of temperature, pH of the water, etc. This alteration, combined with in situ mechanical stresses, can lead to mechanical damage that could compromise the product's behavior. Chapter 6 looks at the mechanisms underlying the alteration of geosynthetics that are subject to these physicochemical conditions. To limit the aging of the geotextile and thereby ensure its durability, it is useful to identify all actions to which the geotextile may be subject, in order to select the appropriate product, and also to limit the impact of these actions (e.g. recommending quickly covering the product in order to prevent the geotextile being exposed to UV rays for a long time).

5.3.4.3. *Physical clogging*

Section 5.3.1 introduced the mechanism that makes it possible to obtain a soil–geotextile filtration system that is durable over time, that is, not subject to clogging. Section 5.3.3 laid out the criteria that could allow the creation of this filtration system. However, clogging can still occur in a filtration system comprising a transition zone with the self-filtering soil and/or the geotextile filter, in cases where the geotextile filter cannot remove the unstable, fine particles transported by the flow. In this case, the fine particles may be deposited in front of the geotextile filter, covering the surface (blinding), or may block the openings in the filter, creating a localized clogging (blocking), or may stay trapped in the filter pores, resulting in internal blockage (clogging). It must be noted that the generic term "clogging" can be used as an umbrella term for several of the mechanisms described above. Giroud (2005) considers that while clogging can occur on the surface of any geotextile filter, *blocking* only occurs in woven geotextile filters, while internal clogging occurs only on needle-punched, non-woven geotextile filters. Using a theoretical analysis, Giroud (2005) showed that the clogging on the surface (blinding) was the most critical phenomenon with respect to a reduction in the hydraulic permeability of the filtration system.

The filtration system may be considered to be clogged as soon as its hydraulic permeability falls below that of the upstream soil to be filtered. Depending on the structure, this malfunctioning of the filter may have dramatic consequences, as an increase in the interstitial pressure upstream of the filter could cause the soil to become destabilized (increased force due to the rise in the hydraulic gradient and decreased effective stresses and therefore decreased shear strength). Consequently, in addition to the risks related to improper installation, the criterion regarding the non-retention of fine particles is an essential condition to prevent clogging.

There are two essential points that must be fulfilled to avoid the risk of physical clogging:

– Good knowledge of the in situ soil, covering its physical characteristics and boundary conditions, especially mechanical boundary conditions. For example, Gardoni and Palmeira (1998) showed that the geotextile filtration of residual soils that might contain coarse particles (actually made up of clusters of finer particles) could lead to clogging. For these particular soils, it was necessary to use a dispersing agent in order to know their true granulometry and accurately dimension the geotextile filter. Furthermore, performance tests (section 5.3.5.1) are recommended in order to experimentally determine the long-term evolution of the soil-geotextile interface.

– Good knowledge of the turbidity of the water flow towards the filter. Faure and Fry (2004) describe the clogging of geotextile filters installed upstream of a drainage system in the Petit-Saut dam in French Guiana Guyane, because of water laden with clay particles from the erosion of altered clays.

5.3.4.4. *Absence of a self-filtering transition zone*

If the geotextile filter is able to remove unstable, fine particles, a self-filtering transition zone should get set up between the soil to be filtered and the geotextile to ensure the durability of the filtration system (section 5.3.1). This mechanism, which leads to the soil becoming self-filtering, is only possible when the geosynthetic is applied to the soil with sufficient stress. In cases where this interface is not created, resulting from poor soil-geotextile compatibility, there are two possible ways that the filtration system could malfunction:

– When the soil is unstable, the fine particles, eroded through suffusion, will be removed by the filter. This will not only lead to the drainage system getting clogged, but also destabilize the upstream soil. Thus, the stability condition of a soil is essential. Many authors have studied this and proposed criteria related to this condition (e.g. Kenney and Lau 1985; Lafleur et al. 1993; Bathia and Huang 1995; Cazzuffi et al. 2002; Moraci 2010).

– When the soil is stable, the unstable, fine particles, transported by the flow and passing through the filter, could lead to downstream clogging of the filter, that is, within the drainage system.

5.3.4.5. *Biological clogging*

Biological clogging is caused by microbial activity and develops in the form of a biofilm and incrustations in the geotextile pores, or on the surface, in the case of very fine geotextiles such as non-woven thermal-bonded products. This phenomenon takes place in two stages and then repeats itself: bacteria colonize the solid surfaces in the form of a network of biofilms. This is followed by the precipitation of inorganic substances (made up chiefly of S, Ca and Fe) on the bacteria alone. Since these deposits have micrometric thickness, they form a new support for the growth of secondary biofilm networks whose metabolic activity forms new incrustrations and so on, until the porous material is completely clogged up (Giroud 1996).

Microbial activity can occur in soils with high organic matter content, or in the case of flows laden with micro-organisms, such as leachates from piles of non-hazardous waste buried in storage units.

Many studies have been carried out on the clogging of geotextiles used to filter leachates. For example, Colmanetti and Palmeira (2002) used laboratory tests to compare clogging by leachates on sandy filters and on non-woven geotextile filters. Their study showed that despite an 80% reduction in permeability in both geotextile filters, those tested continued to function till the end of the experiments. However, these conclusions must be treated cautiously as the very long-term effect was not studied, and it is possible that a longer percolation time may result in a critical clogging threshold.

Some manufacturers have developed a technique to coat the textile fibers with biocide particles, such as silver, which are meant to stop any biomass from developing on them. Laboratory tests have demonstrated the effectiveness of this treatment (Blond et al. 2012). However, the efficacy of such products in the long term, and in actual operating conditions, is yet to be demonstrated.

5.3.4.6. *Chemical clogging through iron oxides*

This type of clogging is caused by microbial activity catalyzing a chemical reaction that converts ferrous ions (Fe^{2+}) into iron oxide through the following chemical reaction (Mendonca and Ehrlich 2002):

$$2Fe^{2+} + 3H_2O <=> Fe_2O_3 + 6H^+ + 2e^- \quad [5.4]$$

There are other kinds of reactions that result in the precipitation of metallic oxides, for example, the precipitation of manganese oxides. It is essential to anticipate such reactions in structures where geotextile filters may be installed in order to prevent failures in the structure.

Several studies have shown various types of failures. For example, Palmeira and Fannin (2002) highlighted problems in one section of a canal in Brazil following the clogging of the geotextile filter by ferric oxide. Testemale et al. (1999) also observed that filtration geotextiles installed in the Torcy-Vieux dam were impregnated with iron oxide precipitate. According to the authors, this led to increased retention of clayey particles washed and transported in a suspension. Despite this incipient clogging of the geotextile filter, the authors observed that the filtration function was still being performed inasmuch as the granular downstream drain was clear and the residual permeability of the geotextile remained higher than that of the backfill to be filtered. In conclusion, a precise knowledge of the soil to be filtered and the interstitial water is essential, especially with regard to its mineral concentration and its pH.

5.3.4.7. *Chemical clogging through calcification*

While similar to the previous type of chemical clogging, in this case the ion involved is calcium (Ca^{2+}). Depending on the physicochemical conditions in the environment (Ca^{2+} concentration, pH of the water, relative pressure, temperature, etc.), calcite precipitation may occur. This phenomenon is described in section 5.5.2, which describes feedback from experience from geotextiles used in drainage trenches for 18 years.

5.3.5. *Compatibility tests for soil–geotextile combinations*

5.3.5.1. *Introduction*

The criteria for filters, presented in section 5.3.3, are based on values obtained from index tests on geotextile filters, that is, not related to the ground conditions. These help in ensuring that the geotextile–soil filtration system behaves appropriately in the majority of current structures, that they present no risks and are subject to hydraulic stresses that are not harsh. On the other hand, for structures with higher risks (dams, embankments, stabilizing slopes through drainage, etc.) or for critical filtration conditions (unstable soils, soils with low confinement, turbulent regime, cyclic flows, etc.), there are performance tests that make it possible to verify the behavior of the geotextile–soil filtration system over time. Although there are international standards (ISO) and/or ASTM standards for how these tests should be carried out, these are not very widely applied in France.

5.3.5.2. *Gradient ratio test under constant load*

The gradient ratio (GR) test was developed in the early 1970s (CalhKoun 1972) for sandy soils and makes it possible to determine how the permeability of the soil–geotextile filtration system evolves under a unidirectional flow with permanent regime. The objective is to assess any possible evolution toward the system getting clogged. The ASTM D 5101 standard (ASTM International 2017) describes a test method associated with specific test equipment, but other equipment is also described in scientific bibliographies. This test consists of running a water flow with several constant gradients (1–10) through a soil sample that is in contact with a geotextile filter and measuring the hydraulic head over 24 h, both in the vicinity of the geotextile (25 mm from the geotextile) and far from the geotextile (75 mm from the geotextile). The permeability of the entire column is also measured throughout the test. The soil mass passes through the geotextile at the beginning, and during the flow it can be measured, which enables an assessment of the soil's susceptibility to undergo internal erosion. The GR is defined as the ratio between the gradient in the zone close to the textile (from 0 to 25 mm from the geotextile) and the gradient in the zone beyond the geotextile's zone of influence (from 25 to 75 mm from the geotextile). The GR value indicates how the filtration system will behave against possible lessivage or clogging. For example, a GR value smaller than 1 indicates that the filtering zone (geotextile/soil) is more permeable than the soil to be filtered, which could either correspond to soil lessivage or to the proper establishment of the self-filtering zone defined in section 5.3.1. A GR value close to 1 signifies that the presence of the geotextile does not modify water flow conditions in the soil. Conversely, a GR value > 1 indicates an increase in interstitial pressure in the neighborhood of the geotextile, which is indicative of clogging in the neighborhood of the geotextile or within the geotextile. Another term that is seen in the bibliography is *conductivity ratio*, because the GR for the gradients between the zone close to the textile and the zone beyond the geotextile's zone of influence also corresponds to the ratio between the hydraulic conductivity in the soil to be filtered (beyond the geotextile's zone of influence) and the hydraulic conductivity of the soil–geotextile filtration.

Based on an experimental study by Haliburton and Wood (1982), which included GR tests carried out on several types of filtration geotextiles using soil containing increasing proportions of fine particles (silts), it was demonstrated that the geotextile filter–soil systems that reached a GR value of 3 saw a sudden increase in GR values for any additional, small quantity of fine particles. Consequently, Haliburton and Wood (1982) recommended limiting the GR value to 3 to prevent the clogging of the soil–geotextile filter system over the long term.

Many studies have been carried out using this GR test. For example, the study by Palmeira and Fannin (2002) recommended measuring the gradient closer to the geotextile than described in the ASTM standard, while Lafleur et al. (2002) suggested using the GR test to assess the equilibrium conditions for the soil–geotextile system and recommended that the following GR values be used to qualify the system: clogging, if GR > 2, bridging if 0.5 < GR < 2 (the only equilibrium state) and lessivage (erosion) of the soil if GR < 0.5. Also, as stated earlier, an experimental study based on the GR test showed that a normal confining stress on the soil–geotextile system increased soil retention, which corresponds to an increase in the GR value. While this "improved" retention is a favorable factor for a soil presenting the risk of lessivage (low GR value), it may be unfavorable in the case of a soil–geotextile filter system with GR values that are already high (close to 3, according to Haliburton and Wood (1982)). Finally, Moraci (1996) developed a GR test protocol to differentiate between surface clogging and internal clogging.

5.3.5.3. *Hydraulic conductivity ratio test under a constant head*

The *hydraulic conductivity ratio* test under a constant load was developed for the same purpose of assessing the long-term behavior of the soil–geosynthetic filtration system. The ASTM D 5567 standard (ASTM International 2018) describes a standardized operating mode to carry out this test in a flexible-wall permeameter. After the saturation and consolidation of the soil–geotextile system, the hydraulic conductivity of this system is measured over time, enabling us to calculate the ratio between the hydraulic conductivity at time *t*, compared to the initial conductivity of the system. A decrease in this ratio indicates a clogging of the soil–geosynthetic filtration system being tested.

5.3.5.4. *Simulation of a cyclical, bidirectional flow*

Various soil–geotextile compatibility tests have been developed to simulate the effects of cyclical, bidirectional flow conditions which may be caused, for instance, by wave action. This application of geotextile filters corresponds to the concept of "filter combined with a protective layer" as defined by Giroud (1981) (section 5.2.1.1). These specific tests were either developed based on apparatus like that used in the GR test (Fannin and Hameiri 1999), or using new devices (Cazzuffi et al. 1999; Zhao et al. 2000). Cazzuffi et al. (1999) showed, notably, that a cyclical flow in a soil with a stable internal skeleton could destabilize this soil due to an increased hydraulic gradient or a reduction in the normal confining stress. There is also a standard, BS EN ISO 10772 (BSI 2012b), for a standardized turbulent regime, which yields a measurement for the quantity of soil passing through the geotextile filter being tested.

5.3.5.5. Tests for in situ usage conditions

The soil–geotextile compatibility tests that we saw earlier for the filtration function are carried out such that gravity is parallel to the flow direction. In several configurations on-site, the geotextile filter is positioned perpendicular to gravity, for instance, as with the geotextile filters that enclose the granular materials in trench drains. To get as close to these geometric conditions as possible, authors such as Ding et al. (2002) developed a performance test that was more complex than the conventional, one-dimensional tests, and in which the geotextile went around a drain placed in the middle of a soil mass. In the study by Ding et al. (2002), the results obtained from the two-dimensional studies corroborated the results obtained in a conventional one-dimensional cell.

5.3.6. *Going further*

5.3.6.1. *Introduction*

Considering the diversity and complexity of soils in terms of granulometry, nature and particle shapes, considering the complexity of the geotextile filter pores (chiefly needle-punched, non-woven) and considering the hydraulic and mechanical phenomena at play, the mechanisms involved in the filtration of soils by geotextiles are numerous and complex. Consequently, although theoretical approaches and approaches using numerical modeling of phenomena taken from the scientific bibliography have brought in information leading to a better understanding of the processes, at present this has still not replaced the design criteria defined in section 5.3.3, which have been experimentally established and verified.

5.3.6.2. *Porometry*

Porometry refers to the measurement of the dimension and distribution of the voids that exist between the solid section (particles, fibers, etc.) in a material. In the case of geotextiles, which are fibrous materials, the voids may be considered as possible paths for a simple particle (either a three-dimensional sphere or a two-dimensional disc), delimited by constrictions in the form of narrowed sections that prevent the particle from traveling from one void to another. Porometry models developed for non-woven geotextiles by Masounave et al. (1980), Gourc (1982) and Faure et al. (1990) yield distribution curves for pores in a non-woven geotextile based on cross-sections perpendicular to the plane of the geotextile or sections within the plane of the geotextile. From an experimental point of view, image analysis systems were developed, for instance, by Masounave et al. (1980), to directly measure the distance between fibers on geotextile samples. This type of measurement has the limitation of being a two-dimensional measurement in a three-dimensional system. Finally, a method to determine the size distribution of pores through capillary flow (also called the bubble point method) was developed. This method was based on the principle of enforcing a difference in air-pressure on either side of a geotextile test sample, which has been saturated by a wet fluid beforehand. The increased difference in air pressure resulted in the drainage of the saturated pores whose size was proportional to the pressure difference according to Jurin's law, which is a combination of the Young–Laplace equations and hydrostatic laws. This method also made it possible to determine the filtration opening size and gain more precise information on the geotextile filter (uniform distribution or wide distribution of voids).

All theoretical or experimental methods that aim to assess porometry are relevant for non-woven, needle-punched geotextiles. However, they are not very relevant for non-woven thermal-bonded geotextiles, woven geotextiles, or knitted geotextiles. A two-dimensional point of view makes more sense for these kinds of geotextiles. Consequently, while determining porometry is useful for theoretical considerations and for an understanding of mechanisms, it is not used at present for design purposes.

5.3.6.3. *Impact of compression/traction on filtration openings*

Some authors, for example, Palmeira and Fannin (2002) and Palmeira and Galvis (2017), have demonstrated the effect of a normal stress on the filtration opening size of needle-punched, non-woven, geotextile filters, especially for limited levels of stress (up to around 50 kPa). Higher levels of stress do not have much influence on the size distribution of the openings. In the CFG recommendations guide (Comité français des géosynthétiques 2014), the effect of stress is taken into consideration by a safety coefficient of 3 on the permeability criterion alone. This approach is incomplete however, as the phenomenon is not taken into account with respect to the retention criterion. Choosing a geotextile filter whose filtration opening size may be close to the non-retention of fine particles criterion could induce a risk of clogging if the normal stress on the filter results in the geotextile's filtration opening size decreasing.

As concerns tensile forces within the plane of the geotextile, Moo-Young and Ochola (1999) have shown that the filtration opening sizes of woven geotextiles may increase by a ratio of 3 for tensile deformations from 0 to 8%. However, Moo-Young and Ochola (1999) observed that in non-woven geotextiles the filtration opening sizes were almost unaffected by tensile deformation (for levels of deformation up to 8%). This observation indicates that the effect of a possible tensile strain must be considered in design woven geotextile filters. In the case of biaxial tensile forces, Fourie and Addis (1996) observed a reduction in the filtration opening size for woven gotextiles.

At present, there is no consensus on taking into account any compression that is normal to the plane or tensile forces for designing geotextile filters.

5.3.6.4. *Retention ratio*

Lafleur (1999) defined the "retention ratio" parameter in order to describe conditions that could lead to clogging or the lessivage of the soil to be filtered based on the granulometric characteristics of the soil as well as the characteristics of the geotextile filter. This parameter is the ratio between the filtration opening size of the geotextile and a representative diameter of the soil to be filtered, which could be d_{85}, d_{50} or d_{30}, depending on the granulometric curve for the soil. This retention ratio must be compared to the retention criterion defined in section 5.3.3.1. The experimental study carried out by Lafleur (1999) on soils with uniform and wide granulometries, corroborated by results from Lafleur et al. (2002), showed that the retention ratio made it possible to predict whether the soil–geotextile filtration system evolved toward a critical situation (clogging or lessivage) or toward a stable situation. These studies thus confirmed the relevance of using the soil skeleton retention criterion defined in section 5.3.3.1 and the criterion for the non-retention of fine particles defined in section 5.3.3.2.

5.3.6.5. *Theoretical approaches*

Various authors have suggested methods for the design geotextile filters. Compared to retention criteria, which have been empirically determined (section 5.3.3.1), these methods are chiefly based on the self-filtering transition zone defined in section 5.3.1. For example, the approach used by Moraci (1996) uses the granulometric curve of the soil to be filtered, and based on the filtration opening size O_{95} of the desired geotextile, it is possible to estimate the granulometric curve as well as the distribution of pores in the zone of contact of the geotextile filter. Finally, the study uses a statistical geometric method to study the equilibrium corresponding to the state in which the particles of the soil be filtered are retained by the transition zone. Finally, let us also mention the approach used by Aydilek

(2006), which offers a new retention criterion based on a probabilistic, numerical model. This model makes it possible to predict the soil retention upstream of the geotextile filter and the structure of the interface between the soil to be filtered and geotextile, and also enables us to assess the hydraulic conductivity of this interface zone in order to establish a ratio for the clogging.

5.4. Geotextile filtration of a suspension or sludge

As described earlier, this filtration application chiefly aims to separate water from the solid fraction of a sludge consisting of water and particles in a suspension, generally fine particles and in variable quantities. This mainly involves dewatering the suspension. The geotextile may be present in the form of a textile tube or it may be integrated with a retention barrier. This application is radically different from the geotextile filtration of in situ soils for several reasons:

– In the case of the filtration of soils, the water to be drained is exogeneous and unlimited in quantity, which assumes the long-term durability of the filtration system. However, for the filtration of sludge, the water to be drained is from the sludge itself and limited in quantity, which assumes that the filtration system only needs to be effective enough for the quantity of water to be filtered.

– In the case of the filtration of soils, there is close contact between the solid skeleton of the soil from the very first moments of filtration and this guarantees durability, while in the case of sludge, there is no contact between the filter and the solid particles before the filtration process begins.

– In the case of the filtration of soils, the geotextile plays the role of the filter continuously, while in the case of the filtration of sludge, the geotextile functions as a filter for the first few moments of filtration and then a filter cake is formed, which becomes the filtering medium for the suspension, instead of the geotextile, which now only serves as a support for the filter cake.

– In the case of the filtration of soils, the granulometry of the solid fraction may vary over a large range, while in the case of sludge, it tends toward a fine fraction, with fine particles being in a suspension.

– In the case of the filtration of sludge, the granulometry of the solid fraction may vary, artificially, if flocculants are added. These substances cause the fine particles to form agglutinations, promoting floc formation, thereby creating a more permeable filtration cake.

– In the case of using textile tubes to dewater sludge, vibrating devices are sometimes passed below the tube in order to destructure, and cause the sedimentation of the filtration cake; in the filtration of soils, the geotextile filter can usually only be accessed through complete excavation.

Thus, based on these major differences between the filtration of soils and the filtration of sludge, the methodology, design criteria for the geotextile filters and the associated testing protocols will all differ. At present, no recommendations guide and/or standards have any geotextile design methodology for the filtration of sludge that is as developed and unanimously recognized as those for the filtration of soils. Thus, the design is mainly carried out based on performance tests or feedback from experience. There are relatively few standardized tests. One of these is the *hanging bag* test, which consists of passing sludge through a suspended textile bag and measuring the volumes of the liquids and solids that pass through and that are retained (ASTM International 2011). Performance tests have been developed in the context of research on the performance of geotextile filters in dewatering sludge based on a prototype cell with vertical columns (e.g. Bourgès-Gastaud et al. 2014) or with horizontal columns (e.g. Stoltz et al. 2019).

5.5. Feedback from experience from earthwork structures

5.5.1. *Introduction*

In general, risks related to the aging of the filters, whether granular or geotextile filters, are a major engineering challenge that can affect the stability and, consequently, the durability of structures. The critical level of the structure in which the geotextile filter is installed is essential information for the designer, not only so that they can use the appropriate safety coefficients and/or verify the soil–geotextile compatibility through long-term filtration tests, but especially to decide on the nature of the filter: granular materials or geotextile?

Experience shows that granular filters behave well over time, even for soils whose granulometry is a bit tricky with respect to filtration (e.g. poorly graduated soils). These filters have a high adaptation ability given their significant thickness. Despite the many advantages of geotextiles compared to granular filters, there has been less feedback from experience and they have a smaller capacity for adaptation, given their small thickness.

While geotextiles are widely used today in common filtration applications (drainage below a geomembrane, highway drains, drainage trenches, etc.), their use in dam filtration applications (e.g. hydraulic) is limited. In order to make geosynthetic techniques more common in applications like dams, we must continue gathering feedback from positive experiences after years of operation in real conditions. This concluding section looks at feedback from experience in the case of two different structures, after many years of operation.

5.5.2. *Roissard trench drains*

In order to study the performance of geotextile filters over time and, notably, to better understand aging processes related to clogging, a research project was carried out in two phases. In the project, different types of geotextile filters were installed in trench drains in order to evaluate their behavior over time, to be excavated after 18 years in service in order to measure their residual physical, hydraulic and mechanical properties. This research project, carried out over the long term, was the subject of many articles and scientific communications that may be consulted for an exhaustive description of the project and its results (Faure et al. 1994, 1996, 2011; Veylon et al. 2012, 2016; Mériaux et al. 2013).

The experimental sites were trench drains constructed in September 1993, in Trièves, near the Roissard commune in the southern part of the Isère département (38)[1] in France. The Trièves region was an ancient glacial lake in the Quaternary (Würm), with large shale deposits, whose thickness at some points is 250 m. These soils, which are generally sloping and have low permeability and weak mechanical characteristics in the presence of water, are the site of more or less active landslides, whose impact may be catastrophic. For example, a large landslide occurred in 1981 in the Harmalière valley in Sinard, about 10 km away from Roissard. The reader can find more detailed information on the geology of this region and the geotechnical problems encountered here in Giraud et al. (1991).

The experimental set-up comprised six drainage trenches, 2–2.5 m deep. In addition to a control trench with no filter between the granular drainage zone and the clayey soil to be drained, there were five different trenches with different types of geotextile filters: needle-punched, non-woven geotextile with an area density of 130 g.m^{-2}, a geotextile woven from strips, a monofilament geotextile and a drainage geocomposite with a heat-patterned draining core which has a needle-punched, non-woven geotextile wrapped around it with an area density of 233 g/m^{-2}.

The trench drains were tracked using a measuring and auscultation device from 1993 to 1999, and from 2008 to 2011. The set of results and analyses from these periods can be found in the publications cited above. In September 2011, work was carried out on the ground to excavate the geotextile filters after 18 years in service. In addition to the observations on the ground, this operation also made it possible to

1 *Département* is a French administrative division.

collect soil samples and geotextile samples. The physical and mechanical characteristics of these samples were measured in the laboratory. The soil samples and the solid particles present in the pores and/or on the surface of the geotextiles also underwent chemical analysis. Based on all the analyses that were carried out, it was seen that the geotextile filters were subject to calcification on the surface (woven geotextiles) and/or within their pores (needle-punched, non-woven geotextiles), which led to a drop in the performance of the trench drains, without causing complete clogging (which would have resulted in completely cutting off the flow through the filter). Surface clogging (blinding) by fine particles was not observed; if present, it could have led to the hydraulic permeability of the filtration system falling to a level equivalent to the hydraulic permeability of shale (i.e. the order of 10^{-9} m.s^{-1}).

In conclusion, this demonstration showed that geotextile filters could see a decrease in their performance, seen here in a decrease in permeability and not retention capacity. This decrease is explained by physicochemical phenomena, and engineers have few tools to fight against these, as applying safety coefficients does not solve the problem. Indeed, installing more permeable geotextile filters, to prevent clogging caused by various processes, could, at a given moment, result in the use of a product whose opening sizes are too large to appropriately retain the particles of the solid soil skeleton. On the other hand, the Roissard site study showed that although there was a drop in performance, the geotextile filters still functioned in a satisfactory manner in the trench drain. Finally, the study tended to show that geotextile filters are not meant to completely replace granular filters. Looking at the project in its entirety, and in addition to its technical aspects considering, notably, the economic aspects, critical level, lifespan of the structure and the access to the filter in case any repair is required, the choice between granular filters and geosynthetic filters is a key point to consider.

5.5.3. *Valcros dam*

To the best of the author's knowledge, the first time that a non-woven geotextile filter was used in an earthwork dam was during the construction of the Valcros dam in the Vaucluse département (84) in 1970, a time when the term "geotextile" did not yet exist. It was used in place of the traditional sand filter. This usage was the subject of multiple analysis and sampling campaigns, the conclusions of which are presented in the following scientific articles and communications (Giroud et al. 1977; Giroud 1981; Delmas et al. 1992; Faure et al. 1999; Giroud 2010).

The Valcros dam is a homogeneous dam made of silty sand, 30% of whose mass corresponds to particles with a diameter smaller than 0.075 mm. BIDIM®, a non-woven textile with polyester filaments, manufactured at the time by the industrial Rhône Poulenc Textile, was used in two places:

– In the downstream drain: this application used 1,450 m^2 of textile, with an area density of 300 g/m^2, around a gravel drain of 8–13 mm.

– In the upstream protection: this application used a textile of area density 400 g/m^2 which was either not covered or covered by 25-cm blocks. Indeed, at the time, in addition to a third zone made up of these 25-cm blocks resting on 15 cm of 60- to 40-mm gravel, three experimental zones were also constructed. After years of service, only the zone with exposed geotextile had undergone erosion under the geotextile. Thus, it was demonstrated, on a real structure (but without endangering the structure), that a geotextile alone could not provide sufficient protection against wave action and gully erosion. The zone with the exposed geotextile was repaired by reinstalling the same geotextile, covered by the same 25-cm blocks that were placed on the two adjacent zones.

Various investigative studies have shown that:

– The downstream drain behaved satisfactorily, from the very first time the structure was filled. There were suspended particles in the first few days after the water was first filled, and since then the water has run clear, to date. It is quite normal to observe water slightly laden with fine particles during the initial phase of functioning of a geotextile filter, considering the removal of the fine, unstable particles during the formation of the self-filtering zone (section 5.3.1). To further support this analysis, a natural, granular filter was observed upstream of the geotextile (Faure and Fry 2004). Further, the downstream flow that was measured was consistent with the hydraulic permeability of the earthwork dam and no seepage was observed at the foot of the downstream bank. It must be noted that based on a back analysis of the filtration criteria applied to the geotextile filter installed in the dam (these criteria did not exist at the time), Giroud (2010) demonstrated that the good, long-term behavior of the geotextile was explained notably by the filtration characteristics (filtration opening size, hydraulic permeability, porosity and thickness), which were dimensioned correctly with respect to the soil to be filtered.

– Tensile tests on the samples of the non-exposed geotextiles, sampled after 6 and 22 years in service, showed a negligible reduction in tensile strength over time. A major part of the drop observed (10–20%) was due to damage to the fibers during installation (interpretations based on microscopic observations). Faure et al. (1999) remarked that as Heerten (1992) suggested, a mechanical perforation criterion could be added to the conventional design criteria for specific applications.

– There was practically no impact on the hydraulic permeability of the geotextiles from the presence of the solid particles in the pores of the geotextiles. In other words, there was negligible physical clogging, as observed by Delmas et al. (1992).

In conclusion, the Valcros dam is a striking example of the performance and durability of a geotextile filter when it has been well-dimensioned, as demonstrated by many of the studies cited earlier. On the other hand, it must be noted that both the geotextiles installed in the dam remained "accessible" in case of failure. Although this has not been framed in terms of criteria and is not present in the standards and/or recommendation guides, this concept of the geotextile filter being accessible must still be considered in the technical choice between a granular filter and a geotextile filter.

5.6. Conclusion

The use of geotextiles for the filtration of in situ and compacted soils is today a mature technique with over 50 years of experience behind it. Whether for simple situations or complex cases, for example, in unstable soils, any engineer has access to a considerable amount of scientific material, giving them the theoretical and experimental tools to dimension the material so as to ensure the durability of the filtering system. In addition to a design that is adapted to the site conditions, the quality of the geotextile (homogeneity in its characteristics, durability of fibers, etc.) and the installation conditions are also essential for meeting the performance and durability objectives. The effectiveness of the technique has been proven and demonstrated given that there is very little feedback from experience that reveals any malfunctions resulting in failure. The difficulty an engineer faces today is not a lack of availability of knowledge, but keeping in mind all the knowledge related to the technique, with the risk of making facile, quick judgments that do not take into account the subtleties involved, thereby imperiling the proper, long-term functioning of the filter.

The risk of clogging is one of the factors related to aging that must always be verified, because while physical clogging by solid particles may be prevented by appropriate design, engineers have few tools available to prevent clogging from biological and/or physicochemical processes if the site conditions are conducive to these phenomena. Therefore, the engineer must carry out a global reflection and take into account the critical level of the structure and the access to the filtration system in case of failure, and weigh geotextile filters against granular filters, with the goal of maximizing the durability of the filtration system. In the future, engineers will be able to accumulate much more feedback from experience over a longer term from in-service filtration systems, in order to verify their durability over the very long term, or to detect aging phenomena that may occur over timespans that are longer than those currently verified, for example, the physical aging of the geotextile fibers. Engineers can carry out these investigations in collaboration with researchers in order to describe and understand the mechanisms behind their observations. With regard to other geotextile filtering applications, for example, the filtration of liquids with fine particles in suspension, or the dewatering of sludge whose majority constituent is fine clay particles, there are questions that are yet to be investigated, for example, how the type of geotextile may influence the permeability of the cake that inevitably forms on the contact surface of the geotextile. Finally, future research could highlight new applications of geotextile filters, born of new products developed from the inexhaustible imagination of geotextile manufacturers, and the limitless possibilities offered by machines. In parallel, potential new products could lead to a revision of the design rules that were established based on the products available today.

5.7. References

AFNOR (2017). NF G 38-061, Utilisation des géotextiles et produits apparentés – Systèmes de drainage et de filtration – Dimensionnement et éléments de conception, Terminologie. Standard, AFNOR, Paris.

ASTM International (2011). D 7701, Standard Test Method for Determining the Flow Rate of Water and Suspended Solids from a Geotextile Bag. ASTM, West Conshohocken.

ASTM International (2017). D 5101 – 99, Standard Test Method for Measuring the Soil-Geotextile System Clogging Potential by the Gradient Ratio. ASTM, West Conshohocken.

ASTM International (2018). D 5567 – 94, Standard Test Method for Hydraulic Conductivity Ratio (HCR) Testing of Soil/Geotextile System. ASTM, West Conshohocken.

ASTM International (2021). D 4751 – 04, Standard Test Method for Determining Apparent Opening Size of a Geotextile. ASTM, West Conshohocken.

Aydilek, A.H. (2006). A semi-analytical methodology for development of woven geotextile filter selection criteria. *Geosynthetics International*, 13(2), 59–72.

Aydilek, A.H. (2012). *Filters and Drains. Handbook of Geosynthetic Engineering*, 2nd edition. ICE Publishing, London.

Bathia, S.K. and Huang, Q. (1995). Geotextiles for internally stable/unstable soils. *Geosynthetics Intentional*, 2(3), 537–565.

Blond, E., Fourmont, S., Bloquet, C. (2012). Biological clogging resistance of tubular drainage geocomposites in leachate collection layers. In *Eurogeo 5*, Valence.

Bourgès-Gastaud, S., Stoltz, G., Dolez, P., Blond, É., Touze-Foltz, N. (2014). Laboratory device to characterize electrokinetic geocomposites for fluid fine tailings dewatering. *Canadian Geotechnical Journal*, 52, 505–514.

BSI (2000). BS EN 13562, Geotextiles and geotextile-related products. Determination of resistance to penetration by water (hydrostatic pressure test). Standard, BSI, London.

BSI (2012a). BS EN ISO 10776, Geotextiles and geotextile related products: Determination of water permeability characteristics normal to the plane, under load. Standard, BSI, London.

BSI (2012b). BS EN ISO 10772, Geotextiles – Test method for the determination of the filtration behaviour of geotextiles under turbulent water flow conditions. Standard, BSI, London.

BSI (2015a). BS EN ISO 10318-1, Geosynthetics Part 1: Terms and definitions. Standard, BSI, London.

BSI (2015b). BS EN ISO 10318-2, Geosynthetics Part 2: Symbols and pictograms. Standard, BSI, London.

BSI (2019). BS EN ISO 11058, Geotextiles and geotextile-related products: Determination of water permeability characteristics normal to the plane, without load. Standard, BSI, London.

BSI (2020). BS EN ISO 12956, Geotextiles and geotextile-related products: Determination of the characteristic opening size. Standard, BSI, London.

Calhoun, C.C. (1972). Development of design criteria and acceptance specification for plastic filter cloth. Technical Report S-72-7, U.S. Army Engineer Waterways Experiment Station, Vicksburg.

Cazzuffi, D., Mazzucato, A., Moraci, N., Tondello, M. (1999). A new test apparatus for the study of geotextiles behaviour as filters in unsteady flow conditions. *Geotextiles and Géomembranes*, 17, 313–329.

Cazzuffi, D., Mazzucato, A., Moraci, N. (2002). The use of geotextiles as filters in unsteady flow conditions. In *Proceedings of 7th International Conference on Geosynthetics*, Nice.

Cazzuffi, D., Moraci, N., Mandaglio, M.C., Ielo, D. (2016). Evolution in design of geotextile filters. In *Eurogeo 6*, Ljubljana.

Colmanetti, J.P. and Palmeira, E. (2002). A study on geotextile-leachate interaction by large laboratory tests. In *7th ICG*, Nice.

Comité français des géosynthétiques (2014). Recommandation pour l'emploi des géosynthétiques dans les systèmes de drainage et de filtration. CFG, Rueil-Malmaison [Online]. Available at: http://www.cfg.asso.fr/publications/guides-de-recommandations.

Delmas, P., Nancey, A., Faure, Y.H., Farkouh, B. (1992). Long term behaviour of a nonwoven geotextile in a 21 year-old dam. In *Conf. Geofilters*, Karlsruhe.

Dierickx, W. and Myles, B. (1996). Wet sieving as a European EN-standard for determining the characteristic opening size of geotextiles. In *Recent Developments in Geotextile Filters and Prefabricated Drainage Geocomposites*, Bhatia, S.K. and Suits, L.D. (eds). ASTM, West Conshohocken.

Ding, K., Yu, L., Qu, X., Dong, F., Zhou, H. (2002). An investigation on geotextiles as envelopes of subsurface drainage pipes in Ningxia of China. In *7th ICG*, Nice.

Fannin, R.J. and Hameiri, A. (1999). A gradient ratio device for compatibility testing in cyclic flow. *Geosynthtics' 99*, 2, 1033–1042.

Faure, Y.-H. and Fry, J.J. (2004). Rétention et colmatage des géotextiles. In *5ème Rencontres Géosynthétiques Francophones 2003/2004*, Lille.

Faure, Y.-H., Gourc, J.P., Gendrin, P. (1990). Structural study of porometry and filtration: Opening size of geotextiles. In *Geosynthetics: Microstructure and Performance*, Peggs, I. (ed.). ASTM, West Conshohocken.

Faure, Y.-H., Farkouh, B., Mériaux, P. (1994). Long term behaviour of trench drains in the Alps. In *Fifth International Conference on Geotextiles, Geomembranes and Related Products*, Singapore.

Faure, Y.-H., Kehila, Y., Farkouh, B. (1996). Behaviour of experimental in-situ trench drains in varved clays. In *Proceedings Geofilters'96*, Montreal.

Faure, Y.-H., Farkouh, B., Delmas, P., Nancey, A. (1999). Analysis of geotextile filter behaviour after 21 years in Valcros dam. *Geotextiles and Géomembranes*, 17, 353–370.

Faure, Y.-H., Meriaux, P., Verstaevel, M., Bertrand, C. (2011). Vieillissement des géotextiles-filtres dans les ouvrages de drainage – auscultation du dispositif expérimental de Roissard. In *Rencontres Géosynthétiques 2011*, Tours.

Fayoux, D. and Evon, E. (1982). Influence of the fiber size on the filtration characteristics of needledpunched geotextiles. In *Proceedings of 2nd International Conference on Geotextiles*, Las Vegas.

Fourie, A.B. and Addis, P. (1996). Hydrodynamic sieving tests on geotextiles subjected to tensile loads. In *Proc Geofilters'96*, Montreal.

Gardoni, M.G. and Palmeira, E.M. (1998). The performance of a geotextile filter in tropical soil. In *Proc. 6th ICG*, Atlanta.

Giraud, A., Antoine, P., Van Asch, T.W.J., Nieuwenhuis, J.D. (1991). Geotechnical problems caused by glaciolacustrine clays in the French Alps. *Engineering Geology*, 31, 185–195.

Giroud, J.-P. (1981). Designing with geotextiles. *Matériaux et Constructions*, 14(8), 257–272.

Giroud, J.-P. (1984). Geotextiles and geomembranes. *Geotextiles and Geomembranes*, 1(1), 5–40.

Giroud, J.-P. (1996). Granular filters and geotextile filters. *Geo-filters*, 96, 565–680.

Giroud, J.-P. (2005). Quantification of geosynthetic behavior. *Geosynthetics International*, 12(1), 2–27.

Giroud, J.-P. (2010). Development of criteria for geotextile and granular filters. In *9th International Conference on Geosynthetics*, Guarujá.

Giroud, J.-P., Gourc, J.P., Bally, P., Delmas, P. (1977). Comportement d'un textile non-tissé dans un barrage en terre. In *International Conference on the Use of Fabrics in Geotechnics*, Paris.

Gourc, J.-P. (1982). Quelques aspects du comportement des géotextiles en mécanique des sols. PhD Thesis, Université de Grenoble, Grenoble.

Haliburton, T.A. and Wood, P.D. (1982). Evaluation of the U.S. Army Corps of Engineer: Gradient ratio for geotextile performance. In *Proceedings 2nd Conference on Geotextiles*, Las Vegas.

Heerten, G. (1992). A contribution to the improvement of dimensioning analogies for grain. Filters in geotechnical and hydraulic engineering. In *First International Conference On Geo-Filters*, Karlsruhe.

International Geosynthetics Society (2009). Recommended descriptions of geosynthetics functions, geosynthetics terminology, mathematical and graphical symbols.

ISO (2021). ISO/DTR 18228, Design using geosynthetics – Part 3: Filtration. Standard, ISO, Switzerland.

Kenney, T.C. and Lau, D. (1985). Internal stability of granular filters. *Canadian Geotechnical Journal*, 22(2), 215–225.

Lafleur, J. (1999). Selection of geotextiles to filter broadly graded cohesionless soils. *Geotextiles and Géomembranes*, 17, 299–312.

Lafleur, J., Mlynarek, J., Rollin, A.L. (1993). Filter criteria for well graded cohesionless soils. In *Proc. Filters in Geotechnical and Hydraulic Engineering*, Balkema.

Lafleur, J., Francoeur, J., Faure, Y.-H. (2002). Piping, bridging and blinding of geotextiles as evaluated from the gradient ration test. In *7th ICG*, Nice.

Masounave, J., Denis, R., Rollin, A. (1980). Prediction of hydraulic properties of synthetics non-woven fabrics used in geotechnical work. *Canadian Geotechnical Journal*, 17(4), 517–525.

Mendonca, M.B. and Ehrlich, M. (2002). Ochre biofilm formation in geotextile filters – A theoretical and experimental approach. In *7th ICG*, Nice.

Mériaux, P., Veylon, G., Faure, Y.-H., Gallo, R., Nivon, D., Barthe, M., Bertrand, C. (2013). Vieillissement des géosynthétiques-filtres utilisés dans le drainage des ouvrages hydrauliques : enseignements tirés du démantèlement d'un dispositif expérimental ancien. In *Digues maritimes et fluviales de protection contre les submersions*, Royet, P. and Bonelli, S. (eds). Hermes Lavoisier, Paris.

Moo-Young, H. and Ochola, C. (1999). Strain effects on the filtration properties of geotextiles. In *Proc. Geosynthtics'99*, Boston.

Moraci, N. (1996). A new design criterion for geotextiles as filters. In *Proceedings of the Geofilters '96*, Montréal.

Moraci, N. (2010). Geotextile filters: Design, characterization and factors affecting clogging and blinding limit states. In *9th International Conference on Geosynthetics – Geosynthetics: Advanced Solutions for a Challenging World. ICG 2010*, Guarujá.

Palmeira, E. and Fannin, R.J. (2002). Soil geotextiles compatibility in filtration. In *7th ICG*, Nice.

Palmeira, E. and Galvis, H. (2017). Opening sizes and filtration behaviour of nonwoven geo-textiles under confined and partial clogging conditions. *Geosynthetics International*, 24, 125–138.

Stoltz, G., Delmas, P., Barral, C. (2019). Comparison of the behaviour of various geotextiles used in the filtration of clayey sludge: An experimental study. *Geotextiles and Géomembranes*, 47, 230–242.

Testemale, J.P., Parron, C., Royet, P., Faure, Y.H., Lambert, S., Fagon, Y. (1999). Filtre géotextile du barrage de Torcy-Vieux (Saône et Loire) : prélèvements et analyses. *Rencontres Géosynthétiques 99*, 2, 311–318.

Touze-Foltz, N. (2016). Les géosynthétiques, itinéraire d'une recherche partagée. *Sciences Eaux et Territoires*, 18, 6–8.

Veylon, G., Barthe, M., Mériaux, P., Bernard, A., Gallo, R., Faure, Y.-H. (2012). Ageing of geotextiles in drainage systems and field observations. In *5th European Geosynthetics Congress*, Valence.

Veylon, G., Stoltz, G., Mériaux, P., Faure, Y.-H., Touze-Foltz, N. (2016). Performance of geotextile filters after 18 years' service in drainage trenches. *Geotextiles and Géomembranes*, 44, 515–533.

Zhao, Z.K., Chew, S.H., Karunaratne, G.P. (2000). Performance of geotextile filters in revetment under cyclic wave loading. In *Proceedings of the Second Asian Geosynthetics Conference*, Kuala Lumpur.

6

Durability of Geosynthetics

Laetitia VAN SCHOORS and Fabienne FARCAS
MAST-CPDM, Université Gustave Eiffel, Marne-la-Vallée, France

6.1. Introduction

Geosynthetics are products, at least one of whose components is made from a synthetic or natural polymer, in the form of a sheet, a strip or a three-dimensional structure, used in contact with soil and/or other materials in geotechnical and civil engineering applications (ISO 10318). The durability of geosynthetics was (and remains) a key concern for manufacturers, project managers, site managers and the scientific community. For many applications in civil engineering structures, these materials must have a lifespan of over a century. Their durability depends on multiple factors related to the quality of the geosynthetic, the compatibility between the geosynthetic and the environment, and how well best practices have been followed during the design and installation of the structure.

This chapter will look at the effects of the environment on the geosynthetic depending on their chemical nature. More specifically, we look at estimating durability based on the study of the aging of geosynthetics containing synthetic polymers and additives, which are subject to various environmental stresses (oxygen in the air, temperature, photochemical radiation, aqueous medium, etc.).

A polymer is a set of long macromolecular chains made up of one or more repeating motifs, called "monomers". The mechanical properties of geosynthetics increase considerably with the length of the polymer chains and the chemical reactions (e.g. weak bonds) as well as physical reactions (e.g. interlocking) between the chains. These interactions give them a microstructure, characterized by a

Geosynthetics from Yesterday to Today,
coordinated by Pascal VILLARD.

disorganized (amorphous) phase, and/or an organized (crystalline) phase. In the case of synthetic fibers, which are widely used in geosynthetics, the intallation also generates an oriented, amorphous phase (see Figure 6.1).

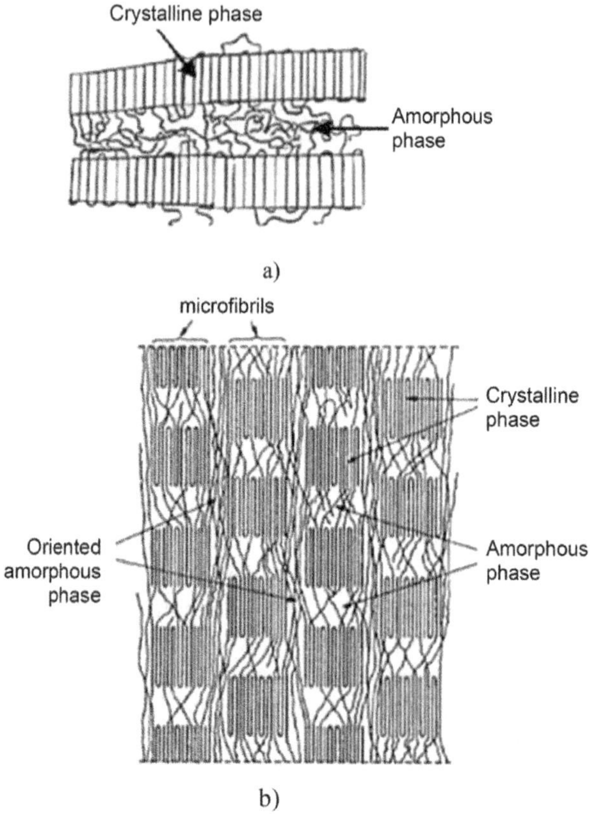

Figure 6.1. *(a) Diagram of the microstructure of a semi-crystalline polymer; (b) microstructural arrangement of a fiber based on Prevorsek's model (Prevorsek et al. 1973)*

The mechanical properties of the polymeric materials are related to the proportion of crystalline phases as well as the quantity and nature of the bonded molecules between the crystalline and amorphous phases.

The aging of geosynthetics depends on the chemical nature of the polymer chains and their microstructure. The crystalline and amorphous phases are not equally accessible to environmental agents that cause their physical and chemical aging. These agents chiefly penetrate into the amorphous phase, while penetrating into the crystalline phase is very difficult, with this phase often being considered to be impermeable (Verdu 1984).

In parallel to the modification of the polymer chain, there are other parameters that influence the aging of geosynthetics. These may be intrinsic parameters like the quantities and specific features of the additives used, and the overall structure of the product, or they may be external parameters, such as the stresses related to the geosynthetic's function and application.

6.2. Main causes for the aging of geosynthetics

Regardless of its nature, the polymeric geosynthetic is placed in a medium that can cause changes in its functional properties over time. These changes are called the "aging of the material". This aging is due to the various environmental stresses related to its use (oxygen in the air, temperature, photochemical radiation, aqueous medium, etc.) and the mechanical stresses that it experiences.

The environmental stresses may cause simultaneous physical and chemical aging, or these may be successive phases, depending on the nature and formulation of the geosynthetic. The rate of aging depends on its thickness and specific area.

6.2.1. *Physical aging*

Physical aging corresponds to a change in the spatial configuration of the polymer chain and/or the overall chemical composition of the geosynthetic, without chemical changes occurring in its constituent macromolecules. This aging may cause changes in the geosynthetic's mechanical properties.

The different types are as follows: structural modifications, the absorption of molecules from the environment surrounding the geosynthetic and migration phenomena (evaporation or the extraction of additives).

6.2.1.1. *Structural changes*

These physical changes are induced by a rise in temperature. They correspond to the rearrangement of polymer chains, which may translate to relaxation phenomena (see Figure 6.2(a)) or anistropic shrinkage and post-crystallization (see Figure 6.2(b)). The post-crystallization generated by the heating of the polymer influences the material's morphology.

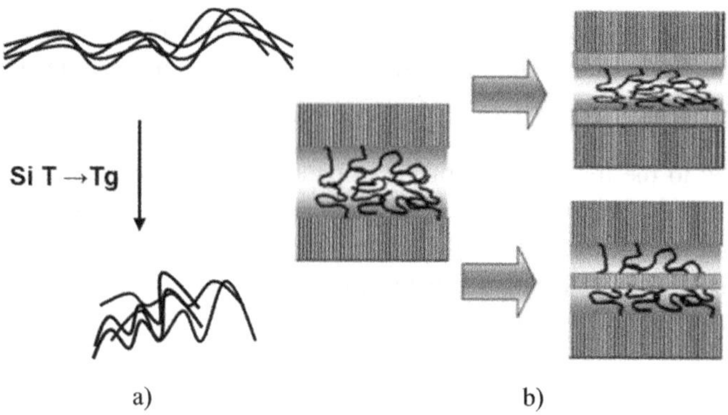

Figure 6.2. *Simplified depiction of the anisotropic shrinkage of materials and the heating process, with (a) either the rearrangement into an amorphous phase on the surface of crystalline lamella or (b) germination through a flaw (Tireau 2011). For a color version of this figure, see www.iste.co.uk/villard/geosynthetics.zip*

6.2.1.2. *Absorption*

Absorption is induced by the diffusion of molecules in the liquid state (aqueous solution) or gaseous state (oxygen) from the surface toward the core of the material, until a balance is reached between the concentrations outside and within the geosynthetic. Diffusion may be compared to molecule transport, which takes place due to a concentration gradient. The diffusion rate depends on parameters intrinsic to the material, such as the chemical nature and the microstructure of the polymer, as well as external parameters such as temperature. It is a thermally activated phenomenon. In a liquid medium, absorption can bring about plastification (the diffusing molecules fill up spaces between the polymer chains) and this could be followed by a swelling of the material.

6.2.1.3. *Migration, evaporation and extraction*

Migration refers to the transport of additives molecules (plasticizers, stabilizers, etc.) from the geosynthetic toward its surrounding environment. As discussed earlier, this is induced through a concentration gradient between the external medium and the geosynthetic. The rate of transport depends on parameters that are intrinsic to the material as well as on environmental conditions. Once the molecules reach the surface of the material, they either evaporate into the air, or are drawn out toward a solid or liquid medium. For example, the migration of plasticizers toward the external medium results in a rigidification of the geosynthetic, which directly impacts its mechanical behavior.

The migration of stabilizers has no immediate effect on the material's functional properties. However, when the stabilizer rate falls below a critical level, the geosynthetic is no longer protected and chemically induced aging may be initiated.

6.2.2. *Chemical aging*

Chemical aging refers to the chemical changes in the polymeric chains that make up the geosynthetic. There may be various causes for this, such as thermo-oxidation, photo-oxidation, radiochemistry, or it may be related to reactions in the surrounding aqueous medium.

6.2.2.1. *Thermo-oxidative aging*

Thermo-oxidative aging takes place in the presence of oxygen and the aging rates increase with the temperature. The reaction kinetics depend on the rate of diffusion of oxygen within the material. Antioxidant stabilizers, which are more reactive with oxygen than the polymers, are added to slow down and inhibit oxidation reactions. The main families of antioxidants used are phenols, phosphits and hindered amine light stabilizers (HALS). Due to their chemical nature, they are oxidized within specific temperature ranges and, thus, protect the polymer during the manufacture, storage, application and usage (see Figure 6.3).

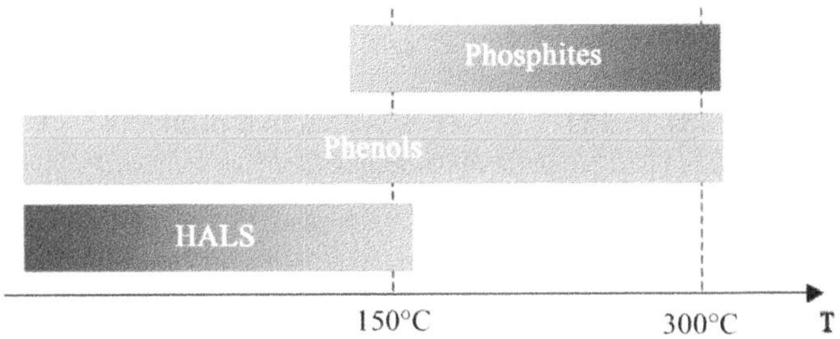

Figure 6.3. *The efficiency of different types of antioxidants with temperature (Gijsman and Sampers 1997)*

The thermo-oxidation reactions in the polymer occur only when the concentration of the stabilizers is too low to protect the polymer from oxidation reactions. Below this critical concentration, the chemical structure of the polymer is modified, which could weaken the material's properties. The thermo-oxidative aging may be represented by three consecutive steps (see Figure 6.4).

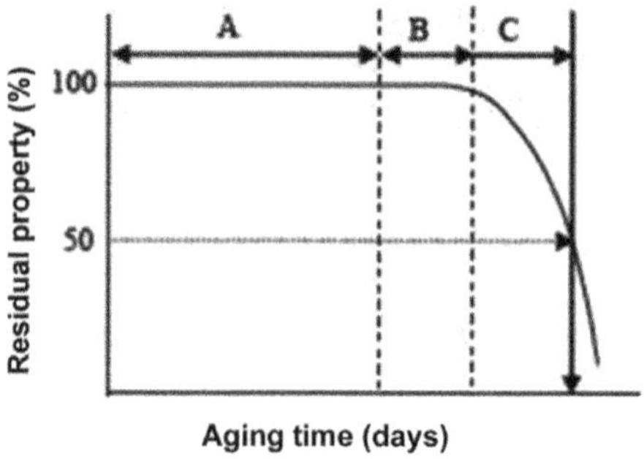

A: depletion time of anti-oxidants (Step 1).

B: induction time to onset of polymer degradation (Step 2).

C: the time to reach 50% degradation of a particular property (Step 3).

Figure 6.4. *The three steps in chemical aging involved in the degradation of an HDPE geomembrane (Hsuan and Koerner 1998)*

The lifespan of polyolefins, defined as per this model, corresponds to the duration of the three steps added. The material reaches its end of life when its loses 50% of the considered property.

Initiation	$\delta POOH$	$\rightarrow$	$\alpha P^{\circ}+\beta PO_2^{\circ}$
Propagation	$P^{\circ} + O_2$	$\rightarrow$	$PO_2^{\circ}PO_2^{\circ}+ PH \quad POOH + P^{\circ}$
Recombination	$PO_2^{\circ} + PO_2^{\circ}$	$\rightarrow$	Inactive products + O_2

With:

– POOH: hydroperoxides

– P: alkyl radical

– PO°: alkoxyl radical

– PO_2°: peroxyl radical

– PH: oxidable site

Figure 6.5. *Thermo-oxidation in the presence of excess oxygen (Bolland and Gee 1946; Colin et al. 2003)*

Thermo-oxidation is a three-step chemical mechanism characterized by a succession of radical chain reactions: initiation, propagation and termination (see Figure 6.5).

This kind of degradation mainly results in the breaking of the chains in the amorphous phase of the polymer, leading to the formation of oxidation products such as ketones, carboxylic acids, esters, alcohols, etc., which cause a reduction in the average molecular weight of the polymer chains. To a lesser extent, these reactions can also form double-bonds, which cause cross-linking (Da Cruz 2015).

6.2.2.2. *Photo-oxidative aging*

Photo-oxidative aging occurs in geosynthetics that are not buried. It takes place in the presence of electromagnetic waves emitted by the sun. The emitted photons are absorbed by the material, causing molecules to shift from their ground state to an excited, electronic state.

The depth at which photons are absorbed depends on the thickness and opacity of the material, but they are generally limited to the surface (a few tens of microns).

Infrared (IR) rays that are part of the solar spectrum influence the photo-oxidation mechanism through an increase in temperature. The combination of UV and IR rays leads to photo-thermal reactions (Verdu 1984).

As in thermo-oxidation, the mechanism of photo-oxidation is made up of a succession of chain radical reactions within the amorphous phase of the polymer. However, unlike in thermo-oxidation, these reactions initially lead to ramification and cross-linking, which result in an increase in the average molecular weight of the polymer chains. In a second phase, the reactions causing the scission of polymer chains predominate, as in thermo-oxidation, and the average molecular weights of the chains decrease.

Various types of stabilizers may be incorporated into the polymer to slow down and inhibit photo-oxidation reactions. The most commonly used stabilizer is black carbon (<3 wt.%). Other kinds of stabilizers that may be used are as follows: UV-absorbent compounds, quenchers of excited states or multifunctional stabilizers.

6.2.2.3. *Radio-oxidative aging*

Radio-oxidative aging occurs in geosynthetics used for nuclear applications. These materials are subject to high-energy ionizing radiations. Unlike photo-oxidation reactions, ionizing radiations have very high energy and affect all chemical bonds in the material, except highly conjugated chemical structures (e.g. aromatic compounds), in which the energy can be distributed over a large number of chemical bonds. Furthermore, these radiations can penetrate within the thickness of the material, both in the amorphous and crystalline phases (Verdu 1984).

The same stabilizers as those used to limit photo-degradation are used in the polymer matrix to limit radio-oxidative aging.

6.2.2.4. *Aging in an aqueous medium*

Aqueous media may be made up of many chemical species that influence the pH (acidic, basic), surface tension (tensio-active, etc.) and ionic composition (salts, metals, etc.).

When geosynthetics are in contact with these media, depending on their chemical nature, they may undergo aging, which occurs through the hydrolysis reactions depicted in Figure 6.6.

$$\text{A-B} + H_2O \rightarrow \text{A-OH} + \text{H-B}$$

Figure 6.6. *Simplified mechanism of the hydrolysis of polymeric chains*

This type of degradation mainly results in the breaking of chains within the amorphous phase of the polymer and, consequently, a reduction in the average molecular weight of the chain.

These four types of aging cause microstructural and morphological changes, related to the chemical nature of the polymer and its environment. All these changes are responsible for variations in the functional properties of a geosynthetic over time.

6.3. Impact of aging on the performance of geosynthetics

In some polymers, changes in mechanical properties cannot be detected early. Thus, a knowledge of physicochemical properties is a way of detecting a state of aging and, in fine, predicting the durability of the geosynthetic.

Since each polymer has a different level of sensitivity to the medium in which it is found, the impact of aging on its performance must be approached based on the polymer's chemical composition.

6.3.1. *Polyolefins*

Polyolefins are materials formed from the polymerization of unsaturated hydrocarbon monomers. The polyolefins that are mainly used in geosynthetics are polypropylene (PP), flexible polypropylene (fPP), polyethylene (PE) and ethylene propylene diene terpolymer (EPDM).

PP is a semi-crystalline thermoplastic polymer used in the form of fibers in geotextiles. fPP is an amorphous thermoplastic copolymer, made up of PP and PE, used in geomembranes. PE is a semi-crystalline thermoplastic polymer that can be synthesized with different densities: High Density PE (HDPE), Medium Density PE (MDPE) and Low Density PE (LDPE). Density is related to the degree of ramification on the main chain – the higher the degree of ramificiation, the lower the density of the polymer. We thus have polymers with different mechanical properties.

PE is used in geomembranes and as envelopes to protect geotextile fibers. MDPE is an elastomer made up of saturated polymer chains composed of ethylene and propylene molecules copolymerized with a diene monomer that allows vulcanization. If the PE content is higher than 65% by mass, the material is semi-crystalline. If not, it is amorphous. MDPE is used in geomembranes.

Polyolefins are sensitive to oxidation phenomena (thermo-, photo- and radio-oxidation). As discussed earlier, antioxidants are added to limit and slow down the oxidation mechanism. Over time, the quantity of these stabilizers decreases either through chemical consumption, or through the process of evaporation or extraction. When the critical concentration is reached, oxidation reactions are initiated and result in an increase in crystallinity that is responsible for the material's behavior changing from initially being ductile to becoming brittle.

Various studies have shown that for thermoplastic polyolefins, the polymer has a critical average molecular weight (M'c) at which this change in mechanical behavior takes place. This average molecular weight is estimated to be 70 ± 30 kg/mol for PE (Fayolle 2009), around 130 kg/mol for PP used for fibers and 200 kg/mol for isotactic PP (Fayolle et al. 2008).

Figure 6.7 illustrates the results obtained by several authors for values for strain at break for PE based on their average molecular weight.

For polyolefin geosynthetics, these degradations led to a drop in the tensile strain at break. The decrease is greater with the increase in oxygen and/or temperature.

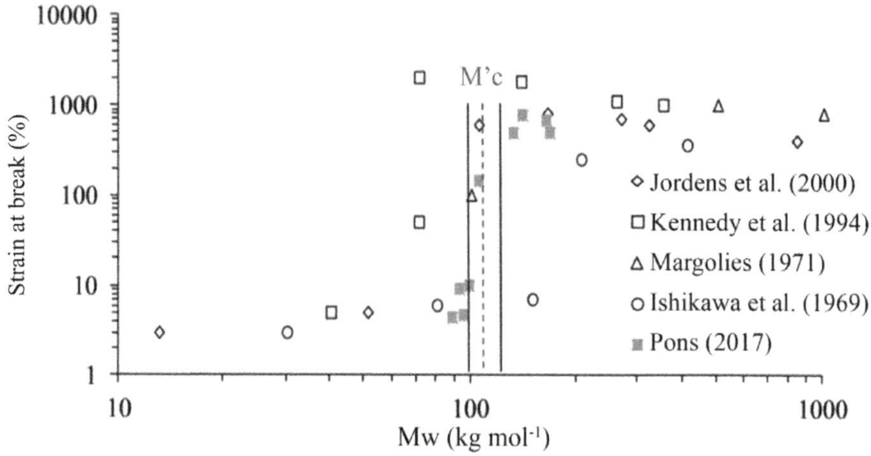

Figure 6.7. *Tensile strain at break as a function of weight average molecular weight of different PEs (Pons 2012). For a color version of this figure, see www.iste.co.uk/villard/geosynthetics.zip*

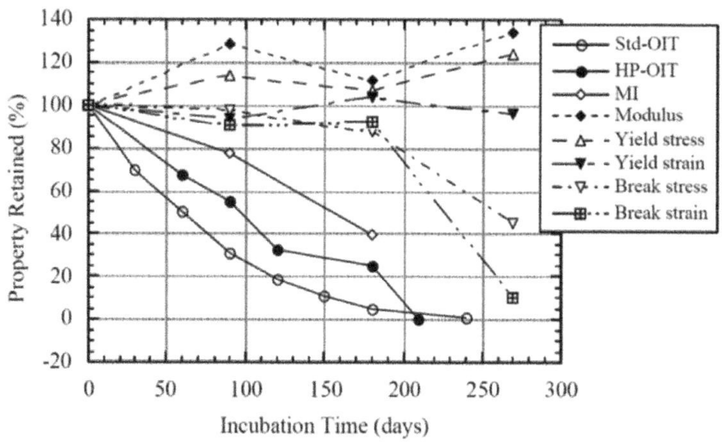

Figure 6.8. *Evaluation of the property retained of HDPE geomembranes up to 360 days of aging in air at 95°C (Hsuan and Guan 1998)*

Hsuan and Guan (1998) showed that in a forced air oven at 115°C, the decrease in tensile strain at break for HDPE geomembranes occurred when the antioxidant concentration was below 5% (see Figure 6.8).

Based on the data obtained during aging in a forced air oven at different temperatures and by applying the Arrhenius equation, the authors estimated that the time required for the antioxidant levels in the HDPE geomembrane to reach a critical concentration was 80 years. In another study, Hsuan and Koerner (1998) predicted a lifespan of 712 years for a geomembrane at a temperature of 20°C, and a lifespan of 109 years at 40°C. Based on the changes in mechanical strength from aging at temperatures between 95°C and 115°C, Greenwood and Curson (2012) predicted a lifespan of 2,700 years at 20°C for PP geogrids. However, establishing and validation of such long lifespans' predictions are debatable.

In some applications, geosynthetics are in contact with aqueous media of different natures (soils with different pH values, water, leachates, etc.), which influence the kinetics of the different steps in the aging process (Rowe and Sangam 2002).

Mueller and Jakob (2003) have shown that stabilizer depletion is faster in water than in air. In one of their examples, the authors did not observe any change in mechanical properties after 13.6 years of exposure of an HDPE to air at 80°C, while at the same temperature in water, they observed a decrease in the properties after five years of immersion.

In the case of PP geotextiles, Richaud et al. (2008) have demonstrated that stabilizers are extracted faster as the alkalinity of the aqueous medium increases.

In the case of HDPE geomembranes immersed in synthetic leachates and representative of leachates from household waste, Sangam and Rowe (2002) showed that tensio-active materials and heavy metals were responsible for the accelerated extraction of antioxidants. This depletion from extraction is likely to occur three to four times faster in leachate than in water. In case of a 2-mm geomembrane, step A, antioxidant depletion was estimated to be 95 years in water and 25 years in a synthetic leachate at 20°C. Pons (2012) showed an acceleration in depletion due to extraction of antioxidants when an HDPE geomembrane is immersed in actual leachate taken from non-hazardous waste storage facilities for household waste, when compared to immersion in water.

PE geomembranes are also subject to stress cracking (Maisonneuve et al. 1997). This is defined as the premature failure of a material resulting from the combined action of mechanical stresses and chemical aggressions related to the environment. This phenomenon can be summarized in four steps (Choi et al. 2007):

– a localized degradation leading to the initiation of multiple microcracks;

– growth of the microcracks;

– the coalescence of microfissures leading to the formation of macrocracks;

– the propagation of fissures deeper into the material, leading to the rupture of the polymer.

Very few studies have been carried out on the aging and durability of MDPE geomembranes.

When used as sealing barriers in holding ponds, Noval et al. (2015) showed that after 21 years of exposure in real conditions of use in the curbing area exposed to the south, there was a small reduction in the thickness of the geomembrane (1.5–1.4 mm) and a drop in mechanical performance. For example, the authors measured an elongation at break of 527% for the geomembrane when new, and an elongation at break of 212% after 21 years of exposure. The shore-A hardness increased from 69, in the initial sample, to 79 after 21 years. These changes were related to a depletion in additives and to oxidation and vulcanization phenomena. On the other hand, flow measurements smaller than 10^{-6} m^3/m^2/day showed that the geomembrane properties that ensured hydraulic performance were still satisfactory after 21 years.

6.3.2. *Polyvinyl chloride*

Polyvinyl chloride (PVC) is produced from the polymerization of monomer units of vinyl chloride ($CH_2 = CHCl$) and is a thermoplastic that is used in many applications, especially as geomembrane for the waterproofing of ponds and dams.

Due to their chemical structure, PVC are rigid materials that require a large quantity of plasticizers, essentially phthalates (25–35%) (Hsuan et al. 2008; Noval et al. 2015), which give them the flexibility needed to their application.

While PVC-P geomembranes have a high chemical resistance to acids, bases, salts and solvents, they are susceptible to oxidation and the extraction of plasticizers and other small molecules. Thus, their aging is caused by a combination of physical phenomena (plasticizer depletion) and chemical phenomena (changes in the polymer chain). This is mainly pointed out through rigidification and change in the material's color.

Rigidification is the result of plasticizer depletion, through extraction in water (Lambert et al. 1999) or through evaporation in air (Wilson 1995), and of chemical reactions that lead to cross-linking (increase in the average molecular weight of the polymer). Regardless of the cause, PVC-P rigidification results in an increase in the glass transition temperature (Tg). The yellow-to-brown coloration is caused by the formation of polyenes.

The aging of PVC through photo-degradation and/or thermo-degradation is a complex process of chemical reactions involving dechlorination (release of HCl), the formation of dienes and a decrease in the average molecular weights of the polymer (see Figure 6.9). These phenomena, induced by the presence of structural faults that appear during polymerization, result in an increased glass transition temperature (Tg) and the material becoming more brittle.

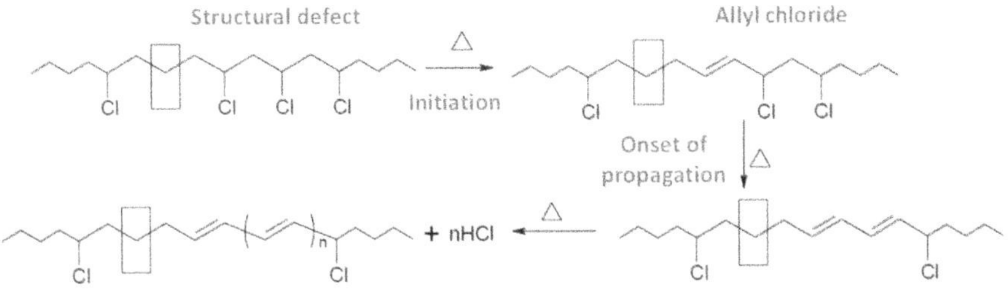

Figure 6.9. *Standard mechanism for PVC dechlorination*

When the PVC-P geomembrane is not exposed to UV radiations, the main limitation regarding its use is plasticizer depletion (Rollin et al. 2002). This is why many studies on long-term performance of these geomembranes have focused on the depletion of these compounds (Morrison and Starbuck 1984; Miller et al. 1991; Giroud and Tisinger 1993; Hammond et al. 1993).

The higher the plasticizer quantity, the more you can delay the aging process (Giroud and Tisinger 1993). However, it is very difficult to produce PVC-P geomembranes with a plasticizer content higher than 35% (Rollin et al. 2002). Taking into account the mobility and volatility of the molecules, Stark et al. (2005) showed that the high molar masses of plasticizers limit their extraction.

Thus, to guarantee that a plasticizer is able to maintain PVC-P performance over time, the FGI 1117 (2019) specifications published by the Fabricated Geomembrane Institute recommend using plasticizers whose molecular weight is at least 400 g/mol.

Another factor that must be taken into account is whether the plasticizers have a linear or ramified structure. Linear pthalates migrate toward liquid environments more easily than ramified pthalates (Giroud and Tisinger 1993; Stark et al. 2005); however, ramified phthalates are more volatile than linear phthalates (Orem and Sears 1979; Wilson 1995). Linear plasticizers are therefore better-suited to external application than ramified plasticizers, especially as linear chemical structures are more photochemically stable. The migration of PVC plasticizers toward water is typically seen in holding ponds, but also takes place when a geomembrane is covered by leachates. Fayoux et al. (1993) showed that the migration of plasticizers in a geomembrane immersed for 10 years in a leachate is slower than in air. Haxo et al. (1985), Artières et al. (1991) and Peggs (1992) showed that PVC geomembranes had good resistance when they are in contact with leachates from municipal waste.

The thickness of the geomembrane also has an impact on aging for physicochemical reasons, as the distance that a plasticizer molecule must travel to migrate outside of the geomembrane increases with the thickness of the geomembrane (Orem and Sears 1979; Rollin et al. 2002; Cazzuffi 2014).

Although the aging of a PVC-P geomembrane is related to the residual plasticizer content, defining its lifespan is rather complicated. In fact, Noval (2015) showed that a 26-year old geomembrane sample from a holding pond in the Canary Islands had plasticizer values lower than 15% (having suffered more than 60% plasticizer depletion) but was still able to carry out its waterproofing function. In a similar way, Blanco et al. (2013) showed that five PVC geomembranes installed in ponds in Spain, which have lost 80% of their plasticizer, continued to carry out their sealing function effectively. Pegg (1992) showed that even if 75% of the plasticizer migrated outside of the geomembrane, the elongation at break remained higher than 100%.

In general, when the formulation is adapted to the purpose for which they are being used, PVP-P geomembranes can still perform their watersealing function, even after 27 years in service (Rollin et al. 2002; Noval 2015).

6.3.3. *Polyethylene terephthalate*

Polyethylene terephthalate (PET) is a product formed by the polycondensation reaction of terephthalic acid and ethylene glycol. This is a semi-crystalline thermoplastic used in the form of fibers in geotextiles, and used as strips in geogrids. High tenacity PET (PET-HT) are materials that offer very good mechanical properties, adapted to soil reinforcement applications.

The PET are sensitive to oxidation reactions (Gijsman et al. 1999) but their aging, when used as geosynthetics, is mainly caused by the action of aqueous solutions that cause physical aging and, most notably, chemical aging.

Numerous authors (Monasse 1990; Bellucci and Nicodemo 1993; Rueda and Varkalis 1995; Bove et al. 1996a, 1996b; Carvalho and Ruvolo Filho 1996; Sammon et al. 2000; Barany et al. 2003) have studied this physical aging, more specifically, swelling, sorption and the diffusion of water molecules within PET films. It appears that water does not cause any significant swelling phenomena in these materials (Sammon et al. 2000).

Chemical aging, and hydrolysis specifically, is the degradation mechanism that governs the durability of PET-HT and results in the breaking of chains at the level of the ester functions (see Figure 6.10).

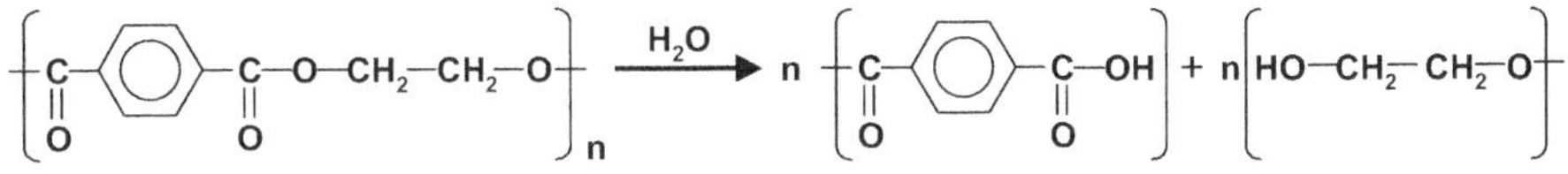

Figure 6.10. *Mechanisms underlying the hydrolysis of PET*

The PET hydrolysis reaction is generally very slow at ambient temperature and the diffusion of water within the polymer matrix is slow. However, the chemical characteristics of the medium may modify the reaction kinetics. Indeed, PET hydrolysis reactions are catalyzed by bases and acids.

Soil pH may be very variable. It can range from 3 for highly acid soils, to over 11 for soils treated with lime or hydraulic binders. These specific features influence not only the rate of hydrolysis but also the localization of the chemical reactions within the material.

In neutral and acidic environments, the hydrolysis reaction mechanisms are identical; however, the lower the pH of the medium, the higher the reaction rate. In these two environments, the degradation of fibers is homogeneous and occurs within the core of the material. It manifests as the scission of chains and, consequently, a decrease in the average molecular weight of the PET. The chain schission leads to changes in mechanical properties.

For example, Risseeuw and Schmidt (1990) showed that when PET-HT is immersed in an acidic medium, if the molecular weight decreased by a factor 3, there was a 50% drop in mechanical strength .

To evaluate the change in performance of PET-HT geosynthetics, various studies (Schmidt et al. 1994; Elias et al. 1998; Schröder et al. 2000; Burgoyne and Merii 2007) produced accelerated aging in water at neutral pH or in an acidic medium.

They showed that in water the hydrolysis reaction was a very slow process and that the mechanical strengths decreased linearly over the aging time. Based on extrapolation methods that used the Arrhenius equation, taking into account either the rate of scission of the chains or the rate of loss of mechanical strength, and regardless of the PET-HT fibers analyzed (Mn > 25,000 g/mol, CEG < 30 meq/g), these authors estimated that the residual mechanical strength after 100 years at 20°C was between 90% and 98%. The different studies showed there was excellent stability in performance over time for these materials when the pH of the medium was close to neutral.

In a highly acidic medium (pH = 1), the estimated residual mechanical strengths after a hundred years at 20°C are of the order of 80% (Elias et al. 1998). Therefore, the mechanical performances are lower than in a neutral medium but remain high for the same duration.

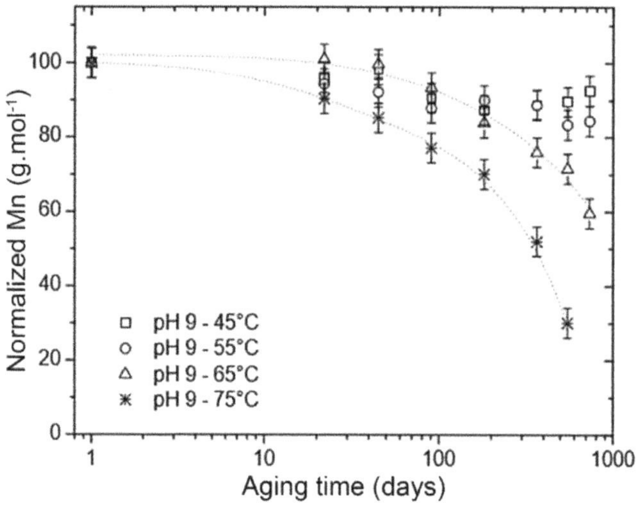

Figure 6.11. *Evolution of the number average molecular weight over the aging time of PET, aged up to 730 days at 45°C, 55°C, 65°C and 75°C at pH 9 (Vouyovitch Van Schoors et al. 2009)*

In alkaline medium, the hydrolysis reactions are catalyzed by OH^- ions. For pH between 7 and less than 10, as with neutral pH, the degradations at the core of the material predominate and result in relatively homogeneous degradation, characterized by a decrease in the average molecular weight of the polymer chains (see Figure 6.11).

For more alkaline environments, that is, for pH values greater than or equal to 10, the hydrolysis is mainly located to the surface. It manifests through the formation of micro-cavities and surface erosion (Halse et al. 1987; Elias et al. 1998; Gu et al. 2001; Vouyovitch Van Schoors et al. 2009). Thus, contrary to homogeneous degradation in acidic or neutral medium, hydrolysis is in this case heterogeneous.

The formation of microcavities on the surface are likely to be linked to the presence of hydrophilic zones on the fiber surfaces. Once these are hydrolyzed, they are likely to become even more hydrophilic and thus lead to the preferential increase of these microcavities (Gu et al. 2001). At these pH levels, it was observed that there was a decrease in the fiber diameters and these shrinkages grew greater as the temperature increased (see Figure 6.12).

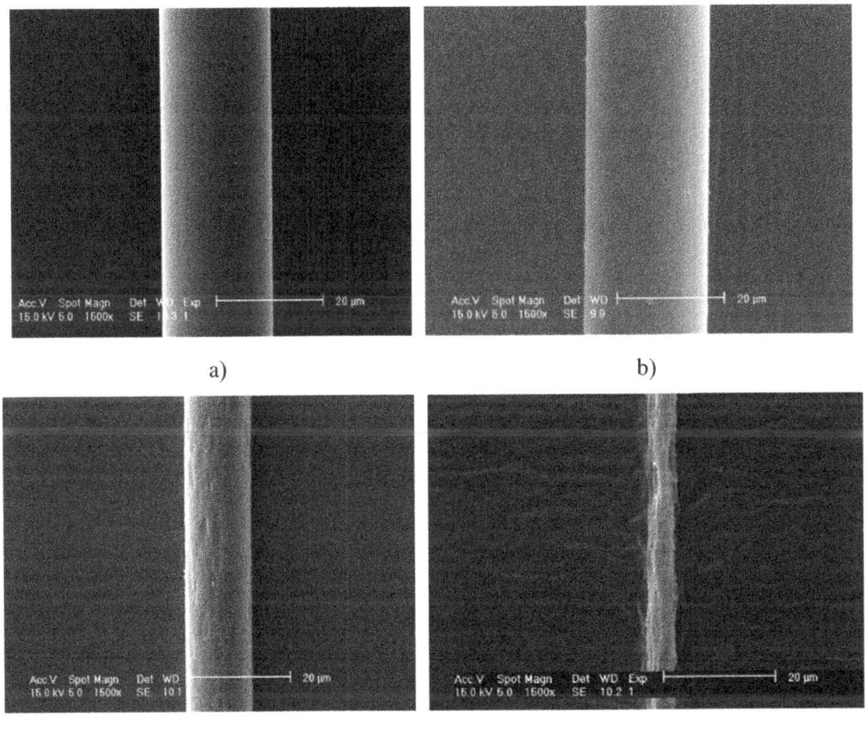

a) b) c) d)

Figure 6.12. *SEM images of PET-HT fibers: (a) not aged; (b) aged; (c) two years at a pH of 11 at 45°C; (d) two years at a pH of 11 and 75°C (Vouyovitch Van Schoors et al. 2009)*

All studies carried out with an alkaline pH showed changes in the mechanical properties that increased with temperature and pH. For a weakly alkaline pH, that is less than 9, the drop in mechanical strength was of the same order as for a neutral pH (Burgoyne and Merii 2007). For pH values between 9 and 10, some studies show changes comparable to immersion in water (Schmidt et al. 1994; Burgoyne and Merii 2007). Other studies, on the contrary, show much larger changes in the mechanical properties of PET-HT than seen for neutral pH. Several authors (Elias et al. 1998; Vouyovitch Van Schoors et al. 2009) evaluated the residual strengths to be of the order of 50–65% over 100 years at 20°C. These contradictory results may be explained by a drop in pH levels over the course of aging, in relation to the carbonation of alkaline solutions by ambient carbon dioxide and the consumption of hydroxyl ions during hydrolysis.

Finally, at very high pH levels, there is a much greater drop in performance. The drop in mechanical strength of PET-HT was estimated to be 1.1% per year at 20°C for a pH value of 11 (Van Schoors et al. 2013) and 4.1% per year for a pH value of 12 (Elias et al. 1998). In these pH conditions, a drop in mechanical strength is mainly due to the shrinking diameter of the fibers (see Figure 6.13).

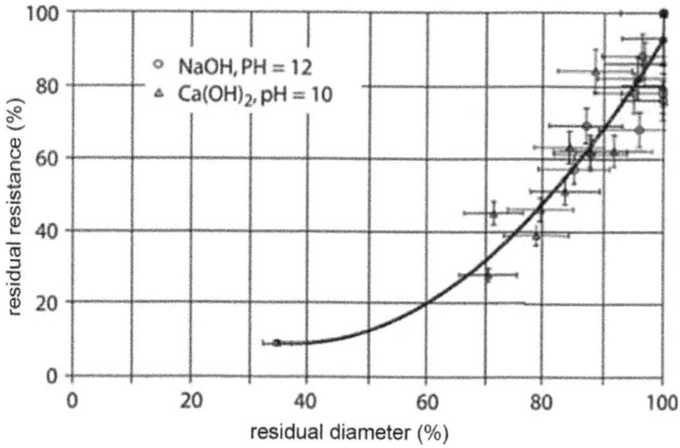

Figure 6.13. *Evolution of the residual strength of PET fibers as a function of the diameter of the PET fibers aged in an alkaline medium at pH values of 10 and 12 (Elias et al. 1998)*

Thus, for applications that require the materials to have a long lifespan, using polyesters in highly alkaline environments would not be appropriate.

6.3.4. *Polyamides and polyaramids*

Polyamides and polyaramids are materials formed from the polycondensation of a di-amine and a di-acid.

In geotextiles, the most commonly used polyamides are aliphatic polyamide 6.6 and aliphatic polyamide 6 (see Figure 6.14).

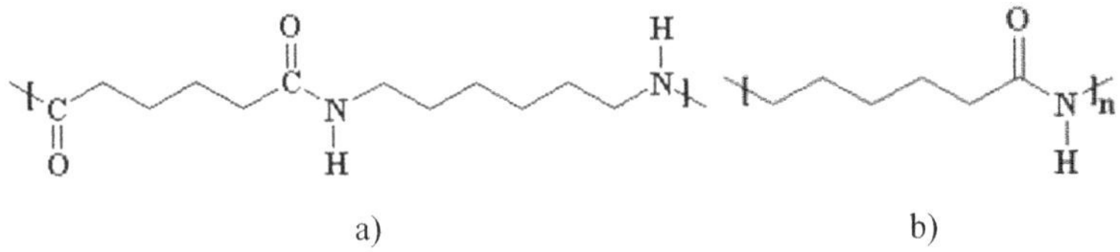

Figure 6.14. *Chemical formula for (a) polyamide 6.6 and (b) polyamide 6*

Polyaramids are aromatic polyamides. There are two main families of polyaramids: poly(p-phenylene terephthalamide) (see Figure 6.15(a)) and copoly (p-phenylene :3,4- diphenyl ether terephthalamide) (see Figure 6.15(b)).

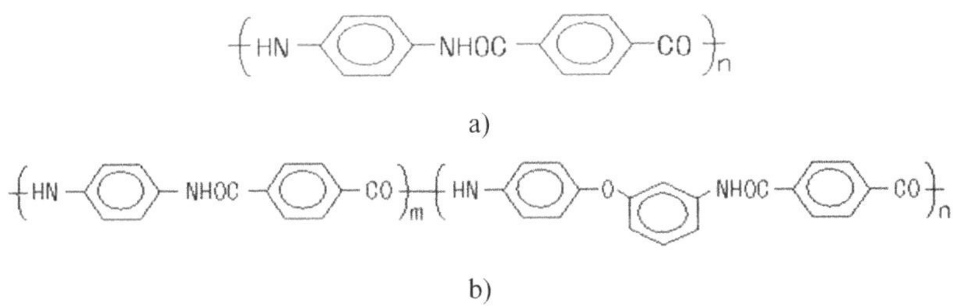

Figure 6.15. *Chemical formula for (a) poly(p-phenylene terephthalamide) and (b) copoly(p-phenylene :3,4- diphenyl ether terephthalamide)*

Polyamide and polyaramid are semi-crystalline polymers, with polyaramids having high crystallinity levels.

Aliphatic polyamides (PA6, PA6-6) used in geosynthetics are sensitive to humidity, oxidation (Dong and Gijsman 2010; Gijsman et al. 2016) and hydrolysis in an acidic medium (Chaupart et al. 1998). They can absorb up to 10% water at 20°C at relatively high humidity levels, or when immersed in water, which causes plastification, and is responsible for decreasing the glass transition temperature (Silva et al. 2013; Arhant et al. 2016). These reversible phenomena lead to swelling and changes in mechanical properties, notably the modulus decreases by a factor 5 for a 9% quantity of absorbed water (Reimschuessel 1978). With respect to irreversible chemical changes during aging at ambient temperature, degradations relate to oxidation predominate over hydrolytic aging (Bernstein and Gillen 2010). As concerns applications other than geosynthetics, for nylon straps, Bernstein and Gillen (2010) estimated that it would take 70 years for these materials to lose 50% of their mechanical strength at 21°C in air. These authors also showed that at ambient temperature, the moisture present in the environment ultimately has very little effect on any drop in the performance of these materials (see Figure 6.16).

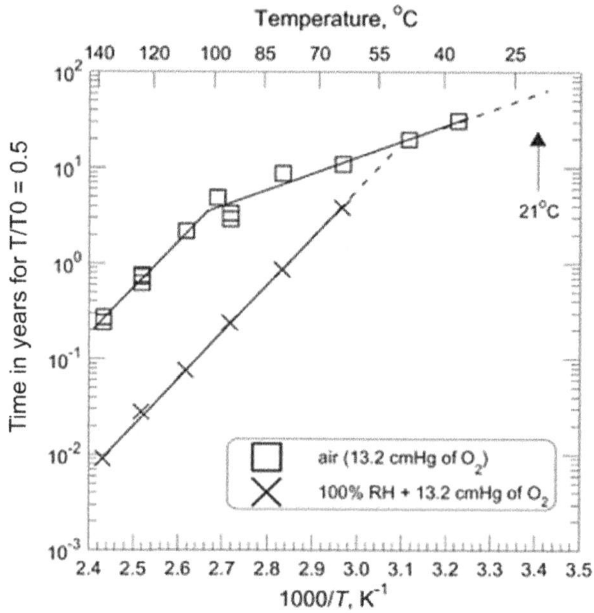

Figure 6.16. *Arrhenius diagram (in years) for the residual tensile strength to become around 50% in air (13.2 cmHg of oxgen) and in the same conditions at 100% relative humidity (Bernstein and Gillen 2010)*

Finally, as stated by Halse et al. (1987), an alkaline medium seems to have little or no effect on the mechanical property of polyamide fibers.

Due to their structure and chemical compositon, polyaramids are likely to be more resistant than polyamides. These materials have an exclusively crystalline structure, which greatly limits the diffusion of aggressive agents during aging. The studies carried out on copoly(p-phenylene:3,4-diphenyl ether terephthalamide) show that after more than 1.5 years of aging at between 20°C and 80°C in deionized water and in alkaline solution with pH 9 and 11, no significant change was observed in the chemical or microstructural characteristics, nor in the mechanical properties, irrespective of the aging conditions (see Table 6.1).

	Temperature	Residual mechanical strength (%)
No aging	–	100 ± 13
pH 11	20°C	101 ± 13
pH 11	80°C	96 ± 10
pH 9	20°C	97 ± 13
pH 9	80°C	96 ± 10
Deionized water	20°C	98 ± 9
Deionized water	80°C	95 ± 10

Table 6.1. *Residual mechanical strength of copoly(p-phenylene:3,4-diphenyl ether terephthalamide) fibers after 1.5 years of aging (Derombise et al. 2009)*

Unlike aliphatic polyamides, poly(p-phenylene terephthalamide) are susceptible to hydrolysis and have little or no susceptibility to oxidation. Studies (Derombise 2009, 2010) carried out over 1.5 years at 20°C, 40°C, 60°C and 80°C, in water and in alkaline solutions at pH 9 and pH 11, reveal degradation phenomena related to the homogeneous hydrolysis of chains at the core of the fibers. At a pH of 9, this hydrolysis leads to the formation of closed porosites. As given in Table 6.2, it appears that the greatest drop in mechanical strength is at 80°C, in the most alkaline media. However, for the same aging conditions, no significant change was observed at 20°C.

	Residual mechanical strength (%)
No aging	100 ± 14
pH 11, 80°C	20 ± 7
pH 9, 80°C	62 ± 20
Deionized water, 80°C	58 ± 7
Air, 80°C	92 ± 12

Table 6.2. *Residual mechanical strength of poly(p-phenylene terephthalamide) fibers after 1.5 years of aging (Derombise et al. 2010)*

Using the Arrhenius equation, the residual mechanical strength is evaluated to be between 81% and 87% after 100 years at 20°C at a pH level of 11. Thus, on the whole, polyaramids have higher durability across these different aging conditions (Derombise 2009).

6.3.5. *Polyvinyl alcohol*

Polyvinyl alcohols (PVA1) are semi-crystalline thermoplastic polymers obtained through the hydrolysis of polyvinyl acetate. High-tenacity polyvinyl alcohol fibers used for geosynthetics are completely hydrolyzed (98%) (Nait-Ali and Freitag 2010). Due to their good mechanical performance, these materials are used for soil reinforcement applications. There have not been many studies on how this type of geosynthetic behaves over time, but the few studies carried out on HT-PVA1 fibers (Sakurada 1985; Nishiyama et al. 2006; Nait-Ali and Freitag 2009; Bian et al. 2018a, 2018b; Bian 2019) showed that these materials are mainly sensitive to humidity and oxidation, especially in an aqueous medium.

Due to their chemical composition, HT-PVAl fibers have a strong affinity for water. In a humid atmosphere (>95°H.R.), the water content in HT-PVA1 yarn is between 5% and 10% (Sakurada 1985; Bian et al. 2018a). This water uptake leads to physical aging and, more specifically, to the plastification of the material, resulting in a decrease in the glass transition temperature and changes in mechanical properties.

For example, at 22°C, when the relative humidity increases from 50% (water content in HT-PVA1 yarn ≈ 1.9%) to 100% (water content in HT-PVA1 yarn ≈ 5.5%), the strain at break of HT-PVA1 yarn increases by about 23% and the secant modulus of elasticity at 2% deformation decreases by about 42%. The mechanical strength remains almost the same. It must be noted that when the same fibers are mechanically constrained, there is less mechanical change. It also appears that in these aging conditions, water does not lead to chemical changes in the fibers (Bian et al. 2018a, 2018b).

With respect to chemical aging, PVA1 seems susceptible to oxidation. Thermo-oxidative aging, carried out between 70°C and 120°C, shows chemical changes related to oxidation and dehydration reactions (Bian 2019). Bian (2019) also showed that these chemical changes induced microstructural changes, notably a decrease in crystallinity and in the melting point of the HT-PVA1 yarn. In these conditions, mechanical changes are observed and are directly related to the changing crystallinity.

Oxidation mechanisms have also been studied in acidic and basic solutions. Some authors (Bian et al. 2018b, 2019) studied the behavior of HT-PVA1 yarn immersed in a solution of sulfuric acid with pH 2.4, at different temperatures, up to 500 days. At 50°C and 60°C, these materials only underwent water-absorption-related aging. In parallel, no chemical changes were observed. On the other hand, at 70°C, the same authors showed that after a year of aging, oxidation occurred, resulting in the formation of unsaturated compounds, esters, ketones and carboxylic acids. These changes were accompanied by a slight decrease in the melting temperature and in the mechanical strength of the fibers (Bian et al. 2018b, 2019). In similar aging conditions (80°C, pH 2, for 48 weeks) and with identical fibers, Nait-Ali et al. (2014) observed no change in the mechanical properties.

According to Bian (2019), the effects of aging in an acidic medium are identical to thermo-oxidative aging, with the acidic solution having an accelerating effect: thermo-oxidative aging at 90°C is equivalent to aging at 70°C in an acidic solution with a pH level of 2.

In an alkaline medium, HT-PVA1 fibers seem to exhibit different behaviors depending on the medium in which aging occurs. There is no significant change in the mechanical properties when aged in a sodium hydroxide solution (pH 12) up to ≈500 days, between 50°C and 70°C (Bian et al. 2018b; Bian 2019) or for 48 weeks at 80°C (Nait-Ali and Freitag 2009). Fibers aged at 80°C, however, undergo a change in color, from white to brown, which is related to oxidation. In a cement extracted solution with a pH level of 12.6, Nishiyama et al. (2006) carried out an accelerated aging test at 20°C and 70°C. As seen earlier, between 20°C and 40°C, there was no change in mechanical strength after three years of aging, despite a slight chemical degradation in the material. At temperatures above 50°C, the higher the aging temperature, the greater the drop in strength. Consequently, after 166 days (4,000 h), the residual strengths were 67% and 42% at 60°C and 70°C, respectively.

All the studies carried out on these fibers tend to show that these materials offer good chemical stability in the temperature conditions usually encountered in soils.

6.4. Conclusion

Various studies have been conducted on the behavior of geosynthetics over time. This has led to significant improvements in knowledge on their aging and durability. As described in this chapter, not all geosynthetics can be used in all environments. Depending on their nature, some are susceptible to oxidation and/or hydrolysis, and/or acids and/or bases, and/or UV radiation, etc.

Defining the characteristics of the environment in which the geosynthetic is used is primordial in selecting the "right geosynthetic", whose lifespan is appropriate for the application. Given the diversity of polymers used in geosynthetics, there is always a solution that is well-suited for short-term, medium-term or long-term use. In fact, the different studies presented in this chapter show that in an environment that is compatible with the considered geosynthetic, the estimated lifespan may be very long (several hundred years).

Although the validity of the prediction of multiple century lifespans is debatable, these studies at least make it possible to demonstrate the high chemical stability of these materials over time. Furthermore, as presented in Chapter 7, structures designed in accordance with best practices do not undergo any particular problems related to the premature degradation of geosynthetics.

Their longevity is being constantly improved, thanks to the requirements related to the products certification, which notably aims to enhance their quality, and also thanks to the standards established for the applications, which have become higher over time and improve the testing methods for durability. While always taking into account the chemical nature of the geosynthetic and using updated knowledge, these standardized tests are carried out at a lower temperature than in the past, therefore reproducing actual aging conditions much better. Due to the lower temperature, the test duration is longer, especially when taking into account service periods that can extend up to a century.

Apart from the chemical nature of the polymers, these tests also look at the depletion of additives like antioxidants and plasticizers. Taking greater account of the various components of the polymer significantly enhances the evaluation of the product's aptitude for service for a specific duration.

However, there is no doubt that tests related to durability must be improved, especially for materials that have not yet been widely studied. We also consider defining simple durability indicators that would make it possible to demonstrate to users how the material's performances are maintained over time.

6.5. References

Arhant, M., Le Gac, P.-Y., Le Gall, M., Burtin, C., Briançon, C., Davies, P. (2016). Modelling the non Fickian water absorption in polyamide 6. *Polymer Degradation and Stability*, 133, 404–412.

Artières, O., Gousse, F., Prigent, E. (1991). Laboratory ageing of geomembranes in municipal landfill leachates. In *Third International Landfill Symposium. Sardinia 91*, Santa Magherita di Pula.

Barany, T., Karger-Kocsis, J., Czigany, T. (2003). Effect of hygrothermal aging on the essential work of fracture response of amorphous poly (ethylene terephthalate) sheets. *Polymer Degradation and Stability*, 82, 271–278.

Bellucci, F. and Nicodemo, L. (1993). Water transport in organic coating. *Corrosion*, 49, 235–247.

Bernstein, R. and Gillen, K.T. (2010). Nylon 6.6 accelerating aging studies: II. Long-term thermal-oxidative and hydrolysis results. *Polymer Degradation and Stability*, 95, 1471–1479.

Bian, Y. (2019). Durabilité des géosynthétiiques en poly (alcool vinylique). PhD Thesis, École Nationale Supérieure d'Arts et Métiers, Paris.

Bian, Y., Colin, X., Aressy, M. (2018a). Polyvinyl alcohol yarns long term properties characterization and assessments. Part I: Physical behavior – Consequences of water affinity. In *11th International Conference on Geosynthetics*, Seoul.

Bian, Y., Colin, X., Aressy, M. (2018b). Polyvinyl alcohol yarns long term properties characterization and assessments. Part II: Chemical durability. In *11th International Conference on Geosynthetics*, Seoul.

Blanco, M., Crespo, M.A., Noval, A.M., Leiro, A., Mateo, B., Aguiar, E., Vara, T., Giroud, J.P. (2013). La durabilidad de geomembranas de PVC-P en función del aditivo uti-lizado como plastificante. In *Congreso Geosintec Iberia 1*, Seville.

Bolland, J.L. and Gee, G. (1946). Kinetic studies in the chemistry of rubber and related materials. III. Thermochemistry and mechanisms of olefin oxidation. *Transactions of the Faraday Society*, 42, 244–252.

Bove, L., D'Aniello, C., Gorrasi, G., Guadagno, L., Vittoria, V. (1996a). Transport properties of dichloromethane in glassy polymers. 6: Poly(ethylene terephthalate). *Journal of Applied Polymer Science*, 62, 1035–1041.

Bove, L., D'Aniello, C., Gorrasi, G., Guadagno, L., Vittoria, V. (1996b). Transport properties of the mesomorphic form of poly(ethylene terephthalate). *Polymer*, 37, 5309–5311.

Burgoyne, C. and Merii, A. (2007). On the hydrolytic stability of polyester yarns. *Journal of Materials Science*, 42, 2867–2878.

Carvalho, G.M. and Ruvolo Filho, A. (1996). Correlation between crystallinity, sorption properties, and some bands in the infrared spectra of poly (ethylene terephthalate) films. *Journal of Macromolecular Science Physics*, B35, 225–264.

Cazzuffi, D. (2014). Long term performance of exposed geomembranes used for the upstream face rehabilitation of dams in Northern Italy. In *10th International Conference on Geosynthetics*, Berlin.

Chaupart, N., Serpe, G., Verdu, J. (1998). Molecular weight distribution and mass changes during polyamide hydrolysis. *Polymer*, 39, 1375–1380.

Choi, B.H.C., Chudnovsky, A., Sehanobish, K. (2007). Letters in fracture and micromechanics, stress corrosion cracking in plastic pipes: Observation and modelling. *International Journal of Fracture*, 145, 81–88.

Colin, X., Fayolle, B., Audouin, L., Verdu, J. (2003). About a quasi-universal character of unstabilised polyethylene thermal oxidation kinetics. *Polymer Degradation and Stability*, 80, 67–74.

Da Cruz, M. (2015). Approche multi-échelle du vieillissement thermo-oxydatif du polyé-thylène utilisé dans les applications de génie civil et de BTP. PhD Thesis, Université Paris Est, Paris.

Derombise, G. (2009). Comportement à long terme des fibres aramides en milieux neutres et alcalins. PhD Thesis, Ecole Nationale des Ponts et Chaussées, Champs-sur-Marne.

Derombise, G., Vouyovitch Van Schoors, L., Davies, P. (2009). Degradation of Technora aramid fibres in alkaline and neutral environments. *Polymer Degradation and Stability*, 94(10), 1615–1620.

Derombise, G., Vouyovitch Van Schoorsm, L., Davies, P. (2010). Degradation of aramid fibers under alkaline and neutral conditions: Relations between the chemical characte-ristics and mechanical properties. *Journal of Applied Polymer Science*, 116, 2504–2514.

Dong, W. and Gijsman, P. (2010). Influence of temperature on the thermo-oxidative degradation of polyamide 6 films. *Polymer Degradation and Stability*, 95, 1054–1062.

Elias, V., Salman, A., Goulias, D. (1998). The effect of pH, resin properties and manufacturing process on laboratory degradation of polyester geotextiles. *Geosynthetics International*, 5, 459–490.

Fayolle, B. (2009). Fragilisation des polymères induite par dégradation, aspects cinétiques. Habilitation à Diriger des Recherches. PhD Thesis, École Nationale Supérieure d'Arts et Métiers, Paris.

Fayolle, B., Richaud, E., Colin, X., Verdu, J. (2008). Degradation-induced embrittlement in semi-crystalline polymers having their amorphous phase in rubbery state. *Journal of Materials Science*, 43, 6999–7012.

Fayoux, D., Gousse, F., Rummens, F. (1993). Assessment on a PVC geomembrane in a landfill after ten years. In *Fourth international landfill symposium*, St Margherita di Pula.

Gijsman, P. and Sampers, J. (1997). The influence of oxygen pressure and temperature on the UV-degradation chemistry of polyethylene. *Polymer Degradation and Stability*, 58, 55–59.

Gijsman, P., Meijers, G., Vitarelli, G. (1999). Comparison of the UV-degradation chemistry of polypropylene, polyethylene, polyamide 6 and polybutylene terephthalate. *Polymer Degradation and Stability*, 65, 433–441.

Gijsman, P., Dong, W., Quintana, A., Celina, M. (2016). Influence of temperature and stabilization on oxygen diffusion limited oxidation profiles of polyamide 6. *Polymer Degradation and Stability*, 130, 83–96.

Giroud, J.P. and Tisinger, L.G. (1993). The influence of plasticizers on the performance of PVC geomembranes. In *Seventh Symposium of the Geosynthetic Research Institute*, Philadelphia.

Greenwood, J.H. and Curson, A. (2012). Life prediction of the oxidation of geogrids by three different methods. *Geotextiles and Geomembranes*, 34, 93–99.

Gu, X., Raghavan, D., Nguyen, T., Vanlandingham, M.R., Yebassa, D. (2001). Characterization of polyester degradation using tapping mode atomic force microscopy: exposure to alkaline solution at room temperature. *Polymer Degradation and Stability*, 74, 139–149.

Halse, Y., Koerner, R.M., Lord Jr, A.E. (1987). Effect of high levels of alkalinity on geotextiles. Part 2: NaOH solution. *Geotextiles and Geomembranes*, 6, 295–305.

Hammond, M., Hsuan, G.Y., Mackey, R.E., Levin, S.B. (1993). The re-examination of a nine-year-old PVC geomembrane used in top cap application. In *30th Annual SWANA Conference*, San Jose.

Haxo, H.E. Jr., Nelson, N.A., Miedema, J.A. (1985). Solubility parameters for predicting membrane-waste liquid compatibility. In *11th Annual Research Symposium: Land Disposal of Hazardous Waste*, Cincinnati.

Hsuan, Y.G. and Guan, Z. (1998) Antioxidant depletion during thermal oxidation of high density polyethylene geomembranes. In *6th International Conference on Geosynthetics*, Atlanta, 375–380.

Hsuan, Y.G. and Koerner, R.M. (1998). Antioxidant depletion lifetime in high density polyethylene geomembranes. *Journal of Geotechnical and Geoenvironmental Engineering*, 124, 532–541.

Hsuan, Y.G., Schroeder, H.F., Rowe, R.K., Muller, W., Greenwood, J., Cazzuffi, D., Koerner, R.M. (2008). Long-term performance and lifetime prediction of geosynthetics. In *4th European Geosynthetics*. Keynote Paper, Edinburgh.

Lambert, S., Duquennoi, C., Tcharkhtchi, A. (1999). Utilisation des géomembranes en altitude : retour d'expérience sur des géomembranes en PVC-P. In *4ème Rencontres géosynthétiques*, Bordeaux.

Maisonneuve, C., Pierson, P., Duquennoi, C., Morin, A. (1997). Accelerated aging tests for geomembranes used in landfills. In *6th International Landfill Symposium*, St Margherita di Pula.

Miller, L.V., Mackey, R.E., Koener, R.M. (1991). Evaluation of a 30-mil PVC liner and leachate collection system I, 10 year old municipal solid waste landfill. In *29th Annual SWANA Conference*, Cincinnati.

Monasse, B. (1990). Les théories de la cristallisation des polymers. *ENSMP*, 15(4/5), 173–200.

Morrison, W.R. and Starbuck, J.G. (1984). Performance of plastics canal lignings. Report, US Department of the Interior, Bureau of Reclamation, Denver.

Mueller, W. and Jakob, I. (2003). Oxidative resistance of high-density polyethylene geomembranes. *Polymer Degradation and Stability*, 79, 161–172.

Nait-Ali, L.K. and Freitag, N. (2009). Renforcement à base de fibres de polyalccol de vinyle : effet des cycle humification/séchage et conséquences sur le dimensionnement dans le cadre des normes françaises. In *7ème Rencontre géosynthétiques*, Nantes.

Nait-Ali, L.K. and Freitag, N. (2010). Reinforcements made of polyvinyl alcohol yarns: Effect of wet/dry cycles and conséquences on the design. In *9th International Conférence on Geosynthetics*, Guarujá.

Nait-Ali, L.K., Thomas, R., Freitag, N. (2014). Polyvinyl alcohol physical behaviour and chemical durability. In *10th International Conference on Geosynthetics*, Berlin.

Nishiyama, M., Yamamoto, R., Hoshiro, H. (2006). Long term durability of Kuralon (PVA fiber) in alkaline condition. In *10th IIBCC International Inorganic Bonded Fibre Composite Conference*, São Paulo.

Noval, A.M. (2015). Estudio de barreras geosintéticas a base de PVC-P, PEAD y EPDM utilizadas en impermeabilización de embalses. PhD Thesis, Université Carlos III, Madrid.

Noval, A.M., Blanco, M., Farcas, F., Touze-Foltz, N. (2015). Etude de la durabilité de la géomembrane en EPDM installée dans la retenue d'eau "el boquerón". In *10èmes Rencontres Géosynthétiques*, La Rochelle.

Orem, J.H. and Sears, J.K. (1979). Flexible poly (vinyl chloride) for long outdoor life. *Journal of Vinyl Technology*, 1, 79–83.

Peggs, I.D. (1992). PVC geomembranes in municipal waste landfill liners and covers: The facts. The PVC Geomembrane Institute, Illinois.

Pons, C. (2012). Durabilité des géomembranes en polyéthylène haute densité utilisées dans les installations de stockage de déchets non dangereux. PhD Thesis, Université Paris Est Marne la Vallée, Paris.

Prevorsek, D.C., Harget, P.J., Sharma, R.K., Reimschuessel, A.C. (1973). Nylon 6 fibers: Changes in structure between moderate and high draw ratios. *Journal of Macromolecular Science*, 8, 127–156.

Reimschuessel, H.K. (1978). Relationships on the effect of water on glass transition temperature and young's modulus of nylon 6. *Journal of Polymer Science: Polymer Chemistry Edition*, 16, 1229–1236.

Richaud, E., Farcas, F., Bartoloméo, P., Fayolle, B., Audouin, L., Verdu, J. (2008). Accelerated ageing of polypropylene geotextiles, the effect of temperature, oxygen pressure and aqueous media on fibers – Methodological aspects. *Geotextiles and Geomembranes*, 26, 71–81.

Risseeuw, P. and Schmidt, H.M. (1990). Hydrolysis of HT polyester yarns in water at mode-rate temperatures. In *4th International Conference on Geotextiles, Geomembranes and Related Products*. The Hague, Brussels.

Rollin, A.L., Pierson, P., Lambert, S. (2002). *Géomembranes – Guide de choix*. Presses Internationales Polytechniques, Montreal.

Rowe, K.R. and Sangam, H.P. (2002). Durability of HDPE geomembranes. *Geotextiles and Geomembranes*, 20, 77–95.

Rueda, D.R. and Varkalis, A. (1995). Water sorption/desorption kinetics in poly(ethylene naphthalene2,6-dicarboxylate) and poly(ethylene terephthalate). *Journal of Polymer Science, Part B: Polymer Physics*, 33, 2263.

Sakurada, I. (1985). *Polyvinyl Alcohol Fibers*. Marcel Dekker, New York.

Sammon, C., Yarwood, J., Everall, N. (2000). A FTIR-ATR study of liquid diffusion processes in PET films: Comparison of water with simple alcohols. *Polymer*, 41, 2521–2534.

Sangam, H.P. and Rowe, R.K. (2002). Effects of exposure conditions on the depletion of antioxidants from high-density polyethylene (HDPE) geomembranes. *Revue canadienne de géotechnique*, 39, 1221–1230.

Schmidt, H.M., Te Pas, F.W.T., Risseeuw, P., Voskamp, W. (1994). The hydrolytic stability of PET yarns under medium alkaline conditions. In *5th International Conference on Geotextiles, Geomembranes and Related Products*, Singapore.

Schröder, H.F., Bahr, H., Kneip, G., Lorenz, E., Schmücking, I., Seifert, S. (2000). Long-term resistance of PET geofibers to inner and outer hydrolysis. In *China International Nonwovens/Techtextiles Conference*, Beijing.

Silva, L., Tognana, S., Salgueiro, W. (2013). Study of the water absorption and its influence on the Young's modulus in a commercial polyamide. *Polymer Testing*, 32, 158–164.

Stark, T.D., Choi, H., Diebel, P.W. (2005). Influence of plasticizer molecular weight on plasticizer retention in PVC geomembranes. *Geosynthetics International*, 1, 99–110.

Tireau, J. (2011). Propriétés à long terme des gaines de polyéthylène haute densité utilisées pour les ponts à haubans. PhD Thesis, École Nationale Supérieure d'Arts et Métiers, Paris.

Van Schoors, L., Nour, I., Moscardelli, S., Barberis, N., Lozach, D., Khay, M. (2013). Comparaison des propriétés des fibres d'un géotextile polyester vieilli en laboratoire et dans des sols traites. In *9ème Rencontre géosynthétiques*, Dijon.

Verdu, J. (1984). *Vieillissement des plastiques*. AFNOR Éditions, Paris.

Vouyovitch Van Schoors, L., Lavaud, S., Duragrin, D., Barberis, N. (2009). Durabilité des géotextiles polyester en milieu modérément alcalin. In *7ème Rencontre géosynthétiques*, Nantes.

Wilson, A.S. (1995). *Plasticizers: Principles and Practice*. The Institute of Materials, CRC Press, Boca Raton.

7

A Review of the Use and Behavior of Geosynthetics in France in the Past 50 Years

Jean-Pierre GOURC[1] and Philippe DELMAS[2]
[1] *Formerly Polytech Grenoble IGE, Université Grenoble Alpes, France*
[2] *Formerly Cnam, Paris, France*

7.1. Relevance of a review of decades-old geosynthetic constructions

The history of geosynthetics in France began in the early 1970s with experiments that fired the imagination of geotechnical engineers of the time: locomotion over soft soil, the construction of embankments on very moist soil (Caen 1970; Puig and Blivet 1973) (Figure 7.1) and the substitution of granular materials in an earth dam (Valcros, see section 7.13) using nonwoven textiles.

All of these were also pioneering applications on an international level.

The development of geosynthetics continued unabated over the following decades, with innovative construction techniques combined which made use of the various functions of geosynthetics.

However, even today, despite the international adoption of geosynthetics in a wide variety of infrastructure applications, these geomaterials are not always chosen for long-term applications. This is because some engineers are skeptical about the longevity of these polymer materials and about how well their ageing is taken into consideration in the design process.

Figure 7.1. *Examples of the pioneering use of nonwoven textiles. For a color version of this figure, see www.iste.co.uk/villard/geosynthetics.zip*

What could be a better demonstration of their longevity than a detailed presentation of construction projects that included geosynthetics and which are still functional three to five decades later?

However, putting together these testimonies based on older constructions, and gathering pertinent information was not an easy task, and it is easy to conclude that without this process, many of these projects would never have been brought to the attention of the new generation of geotechnicians.

The projects presented here are all located in French territories to facilitate the gathering of information and also because France undoubtedly offers the oldest and most diversified range of applications of geosynthetics.

The presentation of these older constructions is divided into three parts.

7.1.1. *Concepts and geotechnology*

It will certainly be interesting to review basic concepts and installation modalities, especially as the present-day standardization of construction methods has led to certain technologies being forgotten, despite being relevant. Examples of this are continuous-filament reinforcement or precast support formwork.

7.1.2. *Design and construction*

The projects presented here, generally accompanied by a review of their principal characteristic data, are largely still operational, several decades after construction. However, the design methods have changed. This is a rare chance to test current methods (Eurocodes, French recommendations, etc.) on projects which de facto have proven themselves safe, since they are still operational decades later.

7.1.3. Long-term behavior and lessons learnt from the application

A number of structures were tracked, instrumented and samples were taken over the course of time. Being able to observe their development over such a long period, with the continuity of the function attributed to the geosynthetic, is an invaluable source of information for present-day designers of structures involving geosynthetics.

7.2. Older structures studied

Table 7.1 lists the structures presented here and reviews their geographic coordinates to allow any engineers who wish to do so to visit the sites. In a way, this is a history of the development of geosynthetic applications in France.

Type of structure	Year of construction	Function and type of geosynthetic	Site code area	Geolocalization Longitude (E) Latitude (N)
Earth dam reinforced downstream	1976	Woven PET geotextile as reinforcement	Vallon de Maraval-Pierrefeu (83)	6°13′45″ 43°15′20″
Reinforced retaining structure	1982	Woven PET geotextile as reinforcement	Prapoutel Les 7 Laux (38)	5°59′33″ 45°15′06″
Reinforced retaining structure	1987	Woven PET geotextile reinforcement + Ebal-Lcpc patented external formwork	Bagnères de Luchon-Hospices de France (31)	42°43′16.6″N 0°38′41.0″E
Reinforced retaining structure	1993	Knitted woven PET geotextile reinforcement + Ebal-Lcpc patented external formwork	Autoroute Foix-Tarascon (09)	42°53′02.4″N 1°37′24.4″E
Landslide stabilization	1984	Woven PP geotextile	Lixing (57)	49°09′00.8″N 6°59′28.5″E

Type of Structure	Year of const.	Function and type of geosynthetic	Site code area	Geolocalization Longitude (E) Latitude (N)
Reinforced segmental wall	–	Woven PP geotextile + Loeffel segmental wall	Brides-les-Bains (73)	–
Bridge abutment scour protection	1990	Woven PP geotextile + Loeffel segmental wall	A26 Chalons-Troyes	(48°54′24.3″N 4°17′03.4″E)
Retaining wall	1988	Sand + PET fiber micro-reinforcement	Gif-sur-Yvette (91)	48°41′38.1″N 2°07′26.6″E
Gas tank cover	1992	Sand + PET fiber micro-reinforcement	Frontenex (73460)	6°18′41″ 45°37′47″
Localized collapse zone	1990	Woven PET geotextile preventing sinkholes	Trois Lucs à la Valentine (13)	5°28′29″ 43°18′13″
Drainage trench	1993	Woven/ nonwoven/ composite filters	Roissard (38)	5°38′39″ 44°52′44″
Earthfill dam	1970	Upstream protection Nonwoven geotextile upstream drainage filter	Valcros La Londe-les Maures (83)	6°16′05″ 43°10′09″
Earthfill dam	1979	Upstream bituminous sealing	L'Ospedale-Porto Vecchio (20)	9°12′26″ 41°40′06″
Earthfill dam	1986	Upstream PVC sealing	Aubrac-St Chely d'Aubrac (12)	3°00′41″ 44°36′56″
Canal bank protection	1994	Geosynthetic-concrete mat	Jonage (69)	5°01′34″ 45°47′23″
Industrial water reservoir	1974	Double sealing of butyl + bituminous geomembrane	Jarry-Pont-de-Claix (38)	5°42′45″ 45°07′07″
Mildly radioactive waste storage cover	1991	Bituminous geomembrane covering composite barrier	La Hague (50)	1°51′44″ 49°40′44″

Table 7.1. *Older structures using geosynthetics presented in this chapter*

7.3. Maraval (1976): dam with reinforced downstream wall

This was one of the first permanent earth structures reinforced with geotextiles (Mitchell and Villet 1987). This structure, in the south-east of France, forms part of a fire-fighting water tank in this dry region (Kern 1977; Cassard et al. 1979).

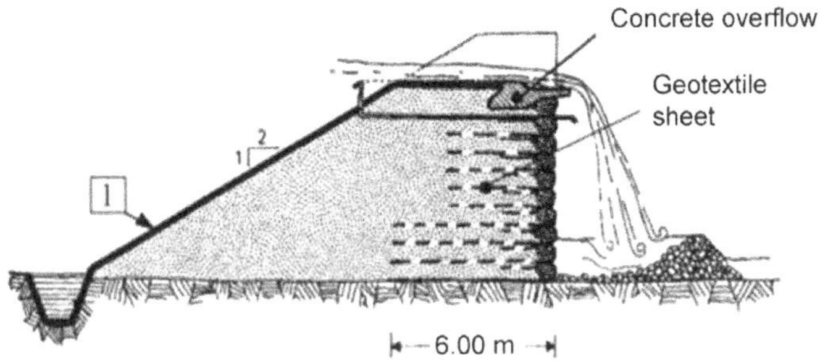

Figure 7.2. *Cross-section of the Maraval dam with its reinforced downstream face*

The downstream face is a nine-meter high vertical wall made of reinforced soil (Figure 7.2). The geotextile used is a woven monofilament polyester (Tri-X) (Figure 7.3). The soil is a mix of clay and compacted schist.

The vertical face is made of gabions filled with gravel round, in which the woven geotextile is wrapped. The woven is impregnated with polymer resin when construction is complete. The spacing between geotextile layers is 0.60 m.

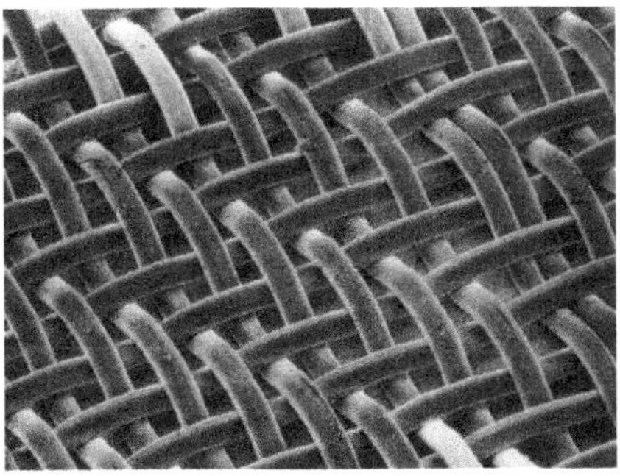

Figure 7.3. *Woven PET used in the Maraval dam*

Compared to its initial appearance in 1976 (Figures 7.4(a) and (b)), the wall showed very small changes over successive visits in 1996 (Figure 7.4(c)) and 2015 (Figure 7.4(d)). The tears observed in 1996, which have not evolved further since then, were most likely caused by animals.

a) Maraval dam - end of construction in 1976

b) Maraval dam - first upstream reservoir impoundment

c) Maraval dam in 1996

d) Maraval dam in 2015

Figure 7.4. *Maraval dam. For a color version of this figure, see www.iste.co.uk/villard/geosynthetics.zip*

7.4. Prapoutel (1982): retaining structure in reinforced soil

7.4.1. *Concept and geotechnology*

This retaining structure with multilayer reinforcement is located at a ski resort near Grenoble (France) at an altitude of around 1,350 m. At the time of its construction, there were very few reinforced retaining structures of similar height in the world.

The chief characteristics of the structure are presented in Table 7.2 and Figure 7.5.

Prapoutel	1982
Geometry	
Height	2–10 m
Length	170 m
Inclination of facing	75°
Geosynthetic	
Type	Woven PET Stabilenka (1)
Tensile strength (warp/weft)	200/50 kN/m (2)
Spacing	1.20 m
Length	4–5 m
Soil	
Angle of friction	$\phi' = 30°$
Cohesion	$c' = 30$ kPa
Unit weight	$\gamma = 18$ kN/m^3
Interface friction angle	$\phi'_{\text{Interface}} = 30°$

(1) Produced at the time by Akzo-Nobel, manufactured by Huesker Synthetic GmbH based on an Akzo-Nobel patent using Enka fibers.

(2) From the manufacturer datasheet (the present-day standard of "tensile strength" had not yet been published).

Table 7.2. *Prapoutel retaining structure with multilayer geotextile reinforcement: geometric and geosynthetic characteristics*

The geosynthetic was a woven geotextile (Figure 7.6). The empirical implementation for the construction of each layer is presented in Figure 7.7. Wooden formwork was used, which facilitated construction of a uniform wall and good control over compaction of the backfill.

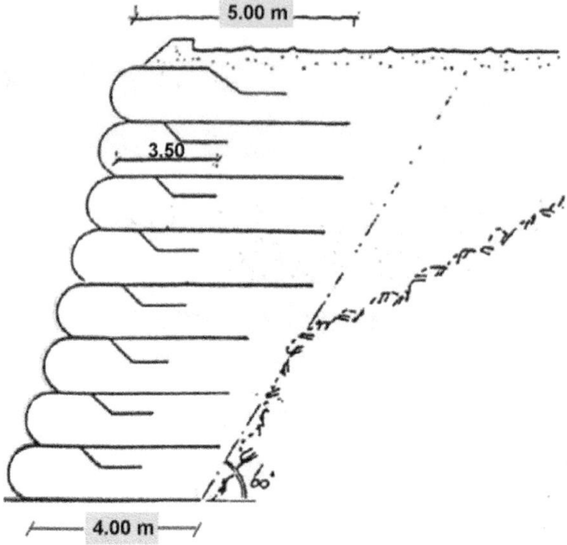

Figure 7.5. *Typical cross-section of the reinforced structure at Prapoutel as constructed in 1982*

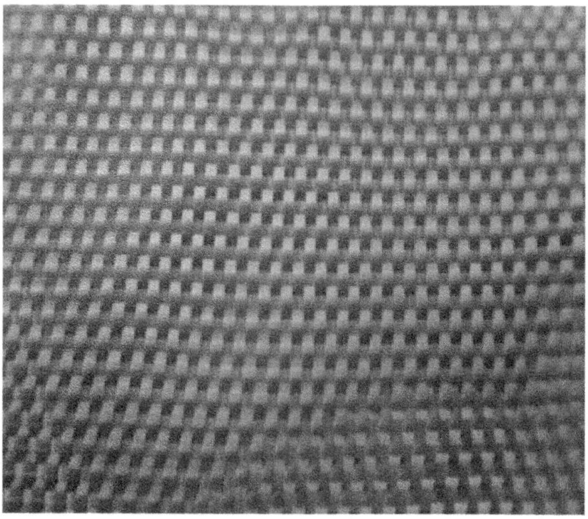

Figure 7.6. *Close-up of the geotextile used to reinforce the Prapoutel support structure (1982)*

Figure 7.7. *Construction details of the reinforced structure at Prapoutel, built in 1982. For a color version of this figure, see www.iste.co.uk/villard/geosynthetics.zip*

7.4.2. *Design and construction*

a) Prapoutel retaining structure during the end of construction in 1982

b) Prapoutel structure in the winter of 1998

Figure 7.8. *View of the Prapoutel retaining structure at different times. For a color version of this figure, see www.iste.co.uk/villard/geosynthetics.zip*

The design approach used at the time was the Broms method (Veldhuizen van Zanten 1986). This structure has two special features (Gourc and Matichard 1994):

– The space between the geotextile layers was 1.20 m, which was quite exceptional at the time (even if it has not rigorously shown to be critical since then; Figure 7.7(a)).

– The facing to protect the geotextile from UV radiation was never built for financial reasons (Figure 7.8). It was already known that geosynthetics are sensitive to exposure to light, but in this case, only a single bitumen layer was applied directly to the exposed surface of the geotextile. The positive side of this is that it is a rare example, allowing us to verify the ageing of geotextiles in aggressive conditions.

7.4.3. *Ageing and lessons learnt from the application*

This is a very interesting case because the project manager's negligence of failing to protect the geotextile face against all regulations already in effect at the time allows us to study the surface degradation due to UV radiation in severe climatic conditions, which are more realistic than any laboratory study.

Until 2008, no damage was observed on the geotextile face.

A sampling campaign was carried out in 1999 (Risseeuw and Alexiew 1999). The main results from the two external samples and one internal sample are presented in Table 7.3.

	Position	Tensile strength (%)	Mw (%)	CEG (meg/kg)	η	Location
Initial value (1982)	–	100	100	26	1.78	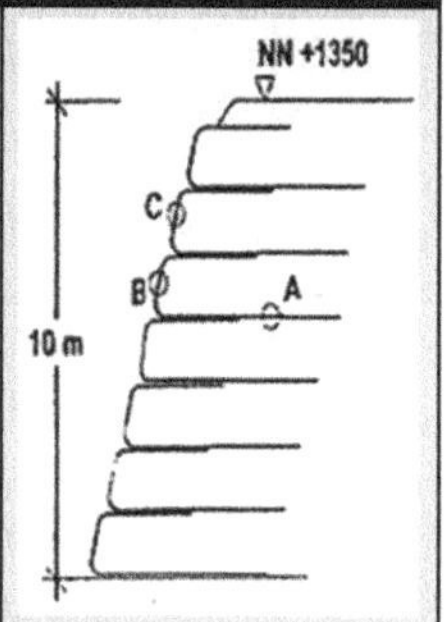
In situ value (1999)	A	95–100	100	26	1.77	
	B	> 25	92	65	1.63	
	C	> 25	86	75	1.63	

Table 7.3. *Changes in the geosynthetic characteristics sampled from the Prapoutel backfill i (September 1997); tensile strength and molecular weight (Mw), concentration of carboxyl end groups (CEG) and relative viscosity (η)*

Degradation due to hydrolysis could be ignored, but due to bad weather (UV degradation), quite logically, a 50% loss of tensile strength was observed on the unprotected south-east face (Risseeuw et al. 1990).

a) Stability of the reinforced embankment structure in Prapoutel in 2009

b) View of the degradation observed on the facing in 2009

Figure 7.9. *Reinforced embankment at Prapoutel in 2009. For a color version of this figure, see www.iste.co.uk/villard/geosynthetics.zip*

The earliest tears in the woven geotextile appeared in 2008, 26 years after construction, mainly at the foot of the embankment (Figure 7.9). It should be noted that no settling or cracking was observed on the surface of the road on top of the embankment, despite local disturbance to the wall face. It therefore seems reasonable to conclude that the ageing affected only the wall face, whose only function is to confine the soil forming the wall face.

In 2013, the owner decided to protect the geotextile wall using a geosynthetic "veil" reinforced by a metallic grid (Figure 7.10).

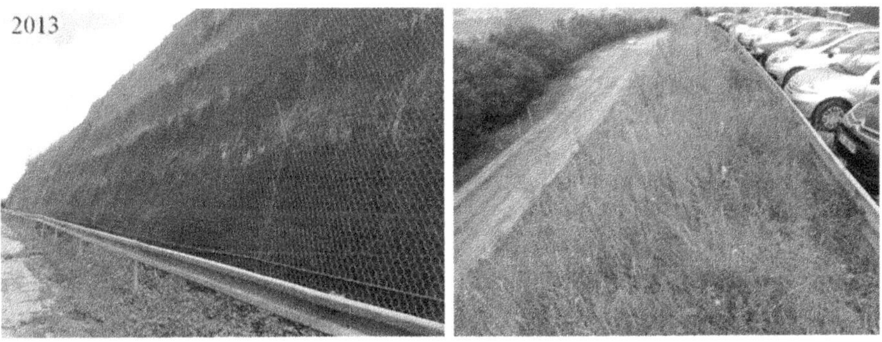

Figure 7.10. *Reinforced backfill at Prapoutel, with its UV protection, in 2013. For a color version of this figure, see www.iste.co.uk/villard/geosynthetics.zip*

7.5. Les Hospices de France (1987): retaining structure

7.5.1. *Concept and geotechnology*

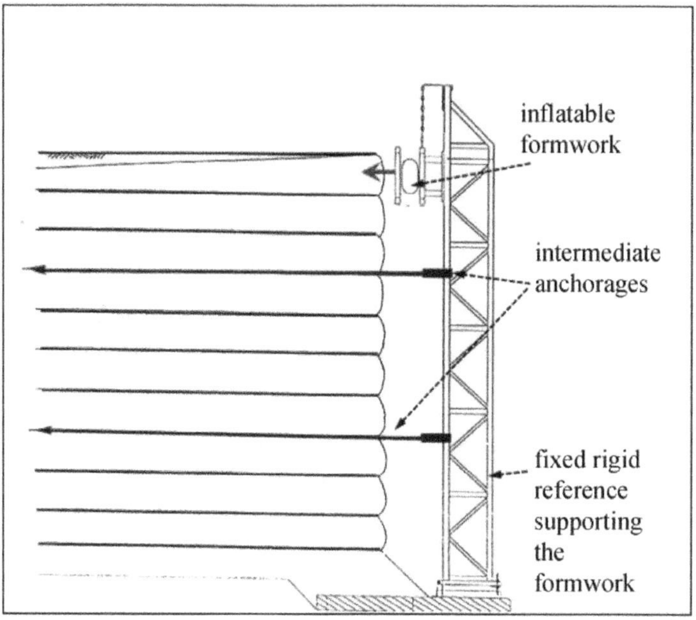

Figure 7.11. *Principle of the EBAL-LCPC concept: the rigid reference that supports the inflatable formwork may be stiffened by intermediate anchors. For a color version of this figure, see www.iste.co.uk/villard/geosynthetics.zip*

The use of geosynthetics as reinforcement in retaining structures has been a well-known technique since the 1970s. Nonetheless, while certain projects, such as the reinforced soil structure at Prapoutel (see section 7.4) demonstrated the feasibility of the concept, the process was not very well developed due to the lack of proven industrial construction techniques (Figures 7.11 and 7.12) and the absence of an aesthetic facing. In 1984, the LCPC (Laboratoire Central des Ponts et des Chaussées (Central Laboratory of Bridges and Roadways)) (Delmas et al. 1984), in collaboration with the company T.P. EBAL, developed a specific technique to facilitate the construction and compacting of a structure built using reinforced soil to offer a solution with an aesthetically pleasing facing.

The following principles were adopted:

– Complete independence of the reinforced structure with the protective surface (Figure 7.12(b)).

– A fixed, rigid reference that temporarily acts as the formwork during the construction. This fixed, rigid reference utilizes either metallic scales (Figures 7.11 and 7.12(a)) or a final facing made using prefabricated elements (Figure 7.12(b)).

– An inflatable formwork system: once inflated, this provides lateral support to the soil layer during compacting (Figures 7.12(c) and (d)). Once deflated, it is easy to remove the formwork and raise it for the layer above (Figure 7.12(a)).

The rigidity and robustness of the system makes it possible to use heavy compactors close to the wall, and it acts as a safety barrier for the workers. At the end of construction, an open space remains between the reinforced structure and the protective facing (Figure 7.13(a)).

This technique has been used in many structures in France, but its fundamental principles have also been used internationally as the basis for the state-of-the-art construction of structures reinforced with geosynthetics (Gourc 1996).

7.5.2. *Design and construction*

The earliest structure built with this technology was the retaining structure on the "Les Hospices de France" roadway, close to Luchon in the Pyrenees. To prevent a landslide like the one that had occurred on the other side of the valley, a new roadway was constructed using geosynthetic reinforcement and the geotechnology techniques presented earlier.

The principal characteristics of the structure are presented in Table 7.4. To facilitate its installation, the geosynthetic was unrolled such that the longer part of the geotextile was on the surface of the soil layer. Figure 7.12(d) depicts the installation of some of the many strain gauges placed on the geotextile at three different levels along four different transversal sections. These were used to study the mobilization of the geosynthetic under load during the construction and post-construction phases. In this case, the inflatable formwork is supported by the prefabricated concrete facing (Figure 7.12(b)).

Figure 7.12. *Phases in the construction of the structure on the "Hospices de France" route. For a color version of this figure, see www.iste.co.uk/villard/geosynthetics.zip*

COMMENT ON FIGURE 7.12. *(a) Metallic scales used to support the inflatable formwork. (b) Concrete protective wall that can also be used as a rigid reference to support the inflatable formwork. (c) Inflatable formwork supported by the prefabricated concrete wall during the construction of a soil layer. (d) Installation of deformation sensor on the reinforcement geotextile.*

A visit to the site in 2016 showed that the structure remained perfectly stable and satisfied the initial specifications from a deformation and aesthetic point of view (Figure 7.13). The space between the geosynthetic enveloping the soil layer and the protective concrete facing still allows us to observe the geosynthetic envelope and does not show any visible degradation in the geosynthetic.

Les Hospices de France	**1987**
Geometry	
Maximum height	6.5 m
Length	84 m
Inclination of the wall	90°
Geosynthetics	
Type	UCO 84464 woven PET (1)
Tensile strength	217 kN/m (2)
Spacing	0.80 m
Length	4 m
Soil	
Particle size	0–200 mm) 70% < 20 mm
c′, ϕ′	10 kPa, 35°

(1) Product produced at the time by UCO.

(2) According to the manufacturer's datasheet (the current tensile strength standard had not yet been published).

Table 7.4. *Les Hospices de France: geometry and characteristics of the geosynthetic and the soil*

7.5.3. *Ageing and lessons learnt from the application*

The wall structure "Les Hospices de France" showed the advantage of intensive compaction, and this was possible thanks to the use of formwork that was supported by a fixed frame.

As can be seen from the deformations recorded in the geotextile layers for two separate heights during construction, more than 80% of the final extension under load (end of construction) occurs immediately after the compaction of the layer (Figure 7.14). This behavior is related to the intensive compaction close to the facing that induces a pretension in the geotextile, which reduces the overall deformations in the wall after construction (Delmas 1988).

a) b)

Figure 7.13. *Hospices de France in 2016: (a) view of the geotextile facing behind the prefabricated concrete wall; (b) view of the prefabricated concrete wall. For a color version of this figure, see www.iste.co.uk/villard/geosynthetics.zip*

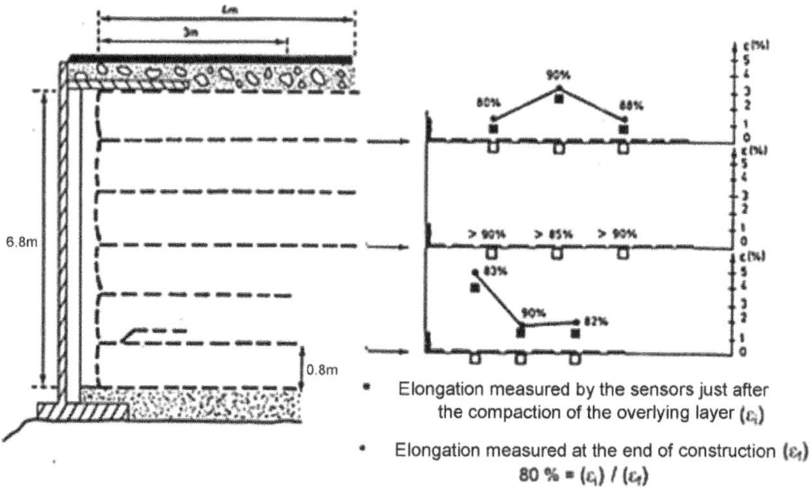

Figure 7.14. *Les Hospices de France: demonstration of the effectiveness of this construction process in inducing a pretension in the geotextile based on the comparative measurement of the extensions of the geotextile after the compaction of the layer, and after construction of the reinforced embankment had been completed*

7.6. Foix-Tarascon highway (1993): retaining structure

7.6.1. *Concept and geotechnology*

As the construction of the reinforced structure on the Hospices de France highway shows, the design given in section 7.5 is an industrial, rather than empirical, approach to the construction of geosynthetic-reinforced structures that use inflatable formwork supported against a fixed-wire mesh anchored support (Figures 7.15 and 7.16). This makes it possible to limit the post-construction deformation of the reinforced structure by pre-stressing the geotextile.

Figure 7.15. *Foix-Tarascon highway-MS13: view of the construction of the third section of the reinforced structure with as the fixed references and inflatable formwork. The protective prefabricated concrete facing has already been placed in front of the lower sections. For a color version of this figure, see www.iste.co.uk/villard/geosynthetics.zip*

At the time, this expanded the field of applications, especially with the possibility of constructing higher structures. This was because the development of new design methods for the geosynthetic-reinforced structure made it possible to account for the specific deformation of the geosynthetics (Delmas et al. 1986; Gourc et al. 1986).

These new design approaches, entitled "displacement methods", took into account an elastic law for the tensile behavior of the geosynthetic and an elastoplastic law for the soil–geosynthetic interface. The distribution of the displacements along the calculated slip surface makes it possible to evaluate the maximum deformation of each layer and the corresponding tensile force mobilized in the geosynthetic.

Figure 7.16. *Foix-Tarascon highway-MS13: view of the third section of finalized reinforced soil. The formworks and scales, which were fixed references, have been removed. The protective concrete wall is being constructed here in front of the last block. For a color version of this figure, see www.iste.co.uk/villard/geosynthetics.zip*

7.6.2. *Design and construction*

The construction of the Foix-Tarascon highway provided an opportunity to build a very high structure: the reinforced MS13 structure presented here (Figure 7.17(a)). The reinforced embankment has a total height of 21 m and is located just above an EDF reservoir (*Electricité de France*: a government-owned French power company). The reinforcement of this structure, which was one of the highest in the world when it was constructed, was provided by warp-knitted geotextiles made using an MDB Texinov process. The principal characteristics of the structure are given in Table 7.5.

The structure consists of three reinforced sections (each 7 m high) and the geosynthetics were designed using the displacement method according to the design standards in use in France in the 1990s (Hérault et al. 1995). The corresponding cross-section is depicted in Figure 7.17(b).

This structure was tracked over time and has shown that the geotechnology adopted at the time made it possible to control deformations in the upper half of the wall because of the use of pre-stressed geotextiles. Logically speaking, the effect of the pre-stressing disappears for the lower part of the wall because of increased loading due to the height of the wall. But the pre-stressing of the geotextiles in the lower part of the structure nevertheless delayed and reduced the anticipate deformations (Blivet and Gourc 1998; Gourc et al. 2001).

Foix-Tarascon highway – MS13 (1993)	Section 1 (lower)	Section 2	Section 3 (upper)
Geometry			
Height	7 m	7 m	7 m
Length	150 m		
Inclination of the wall	90°	90°	90°
Geosynthetics			
Type	BidimRock (1)		
Tensile strength (decreasing toward the top)	400/300 kN/m	200/150/100 kN/m	75/50 kN/mm
Spacing	0.4 m	0.4 m	0.4 m
Length	8.5 m	6.5 m	5/3
Soil			
c′, ϕ′	0 kPa, 35°	0 kPa, 35°	0 kPa, 35°

(1) Produced at the time by Bidim Geosynthetics in 1993.

Table 7.5. *Foix-Tarascon highway-MS13: geosynthetic-reinforced structure, 21 m high: geometric and geosynthetic characteristics*

A visit to the site in 2016 confirmed that the structure was behaving well. As was observed just after the construction and later, water infiltration was observed at certain areas on the geotextile surface. However, no serious damage was noticed at these places (Figure 7.18(b)). As with the reinforced structure on the "Hospices de France" route, the view of the spacing (80 cm) between the geotextile envelope around the layers and the protective, prefabricated concrete panels (here at the foot of the structure, Figure 7.18(a)) and the overall view (Figure 7.17(a)) confirm that the structure was still continuing to fulfill its function well, more than 25 years after construction.

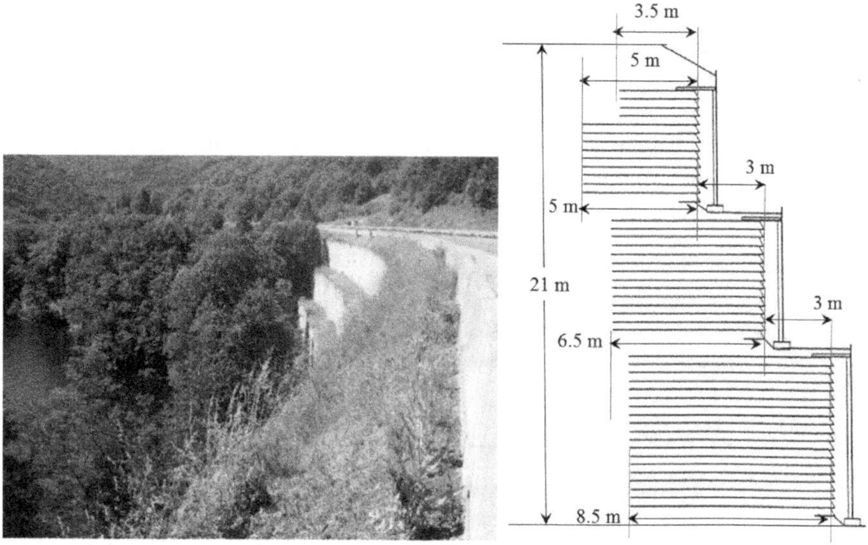

a) b)

Figure 7.17. *Foix-Tarascon highway-MS13: (a) general view of the highway above the reinforced structure (2016); (b) typical cross-section. For a color version of this figure, see www.iste.co.uk/villard/geosynthetics.zip*

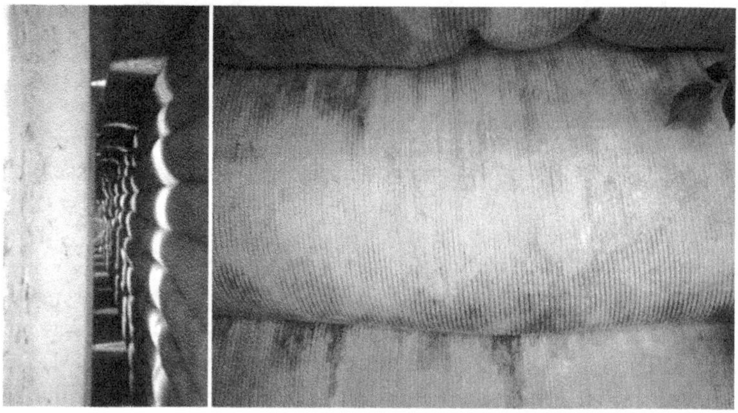

a) b)

Figure 7.18. *Foix-Tarascon highway-MS13. For a color version of this figure, see www.iste.co.uk/villard/geosynthetics.zip*

COMMENT ON FIGURE 7.18.– *(a) View of the spacing (about 80 cm) between the wrap-round on the reinforced structure and the protective concrete panels (2016). (b) View of the geotextile at the level of one of the monitoring windows in the concrete wall at the level of the lower section (2016). As was observed immediately after construction, a small water outflow was visible at certain spots at the level of the horizontal geotextile sheets.*

The maximum tensile strength required, calculated in 1993, was $T_{max.d}$ = 125 kN/m (Table 7.6). Currently, according to the Eurocodes (using the same software based on the "displacement method"), the maximum tensile strength is only $T_{max.d}$ = 99 kN/m. This value was obtained for the same critical circle (Table 7.6) and for the same geosynthetic sheet in the lower part of the structure.

It is also useful to note that since 1993 there has been a slight shift in the approach to the geotextile's tensile strength. This is due to the greater accuracy of data on the base materials and the polymers, with longer testing periods, especially for creep and chemical resistance. The coefficients to be applied to the characteristic tensile strength $R_{t;k}$ to obtain the ultimate tensile strength available, $R_{t.d}$, are also different from the coefficients used in 1993.

Number of layers	Length of the sheets, m	Calculated tensile strength, kN/m	Original slip circles
4	3.5	15	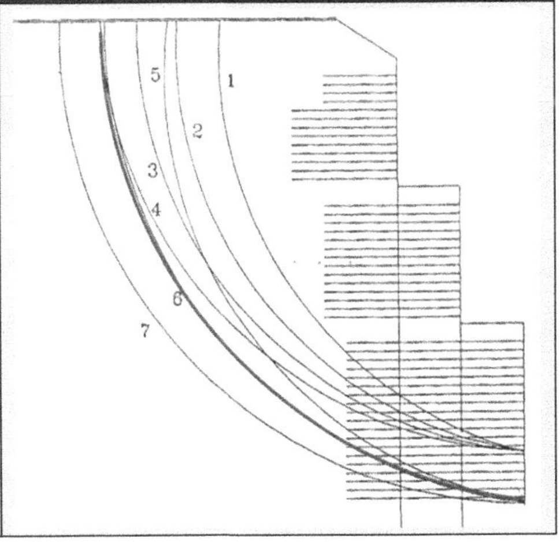
6	5.0	15	
3	5.0	23	
2	6.6	30	
3	6.6	45	
9	6.6	61	
4	8.5	91	
15	8.5	*125*	

Table 7.6. *Original design conditions (1993). For a color version of this table, see www.iste.co.uk/villard/geosynthetics.zip*

Finally, following the French design standards (NF G38-064; AFNOR 2016), the design method used today gives a characteristic minimum tensile-strength value of $R_{t;k}$ = 348 kN/m, lower than the value obtained in 1993 (417 kN/m).

As a reminder, the geosynthetic chosen for the foot of the structure had the characteristic tensile strength $R_{;k}$ = 400 kN/m. In this example, therefore, the traditional design method was more conservative than the Eurocode approach with regard to the characteristic tensile strength. However, it must be noted that the Eurocode approach has placed an additional requirement on the geosynthetic (computation at the Serviceability Limit State) by enforcing a maximum elongation of 3% between the end of construction and the required lifespan of the structure. Chapter 2 gives a detailed comparison of these designs, taking into account the methods, the reduction coefficients and the partial coefficients of 1993 and those of 2023 (section 2.4.5).

7.6.3. *Ageing and lessons learnt from the application*

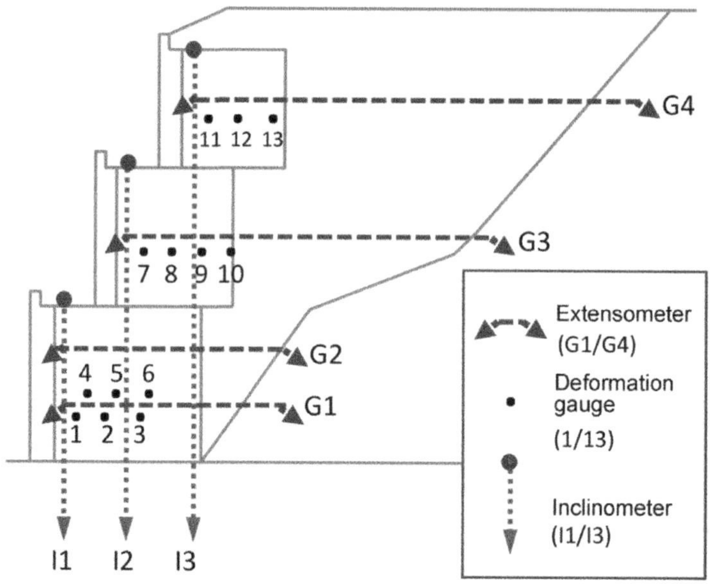

Figure 7.19. *Foix-Tarascon highway-MS13: view of the placement of different instrumentation devices installed in 1993*

It is interesting to note that the measurement campaign that began after construction has confirmed that the deformations stabilized. Figure 7.19 depicts the typical instrumentation of two instrumented cross-sections. This included four horizontal extensometers, fixed to the rock surface behind the structure, strain gauges fixed onto the geotextile for each section, and three vertical inclinometers on each level of the berm. Figure 7.20 depicts the evolution of the horizontal displacements during and just after the construction, measured by the horizontal extensometers.

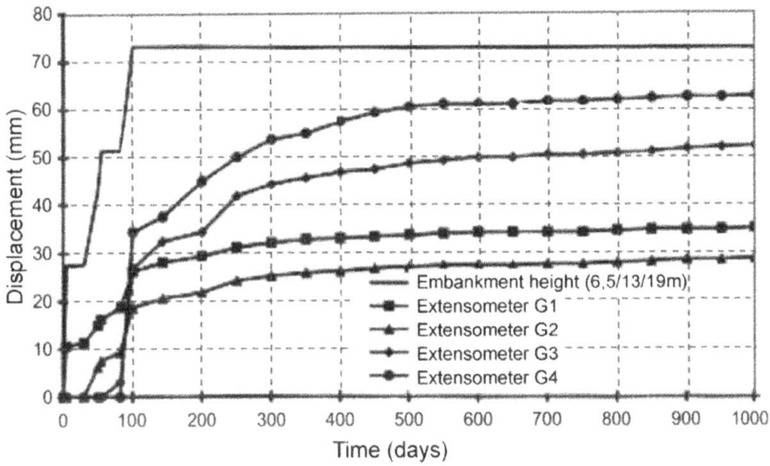

Figure 7.20. *Foix-Tarascon highway-MS13: horizontal displacements during and immediately after the construction, evaluated using the horizontal extensometers*

7.7. Lixing (1984): landslide stabilization using reinforced structures

7.7.1. *Concept and geotechnology*

Stabilizing landslides is one of the trickiest problems for the geotechnical engineer to solve. This is because of the magnitude of the forces that must be mobilized to stabilize the movements and also because the buttressing solution must accept the deformations induced by the unstable mass. Especially when conventional draining solutions are precluded by specific hydrogeological conditions, the use of massive, geosynthetic-reinforced retaining structures may be an interesting alternative.

In order to assess the horizontal forces induced by the sliding mass (RH) at the top of the slip, it is possible to use, for example (Figure 7.21), the approach proposed by Cartier (1985) where:

$$\mathrm{RH}(x) = \sum_0^x (N_i.\sin\alpha_i - T_i\cos\alpha_i)$$

where Ni, Ti and α_i are as defined in Figure 7.21.

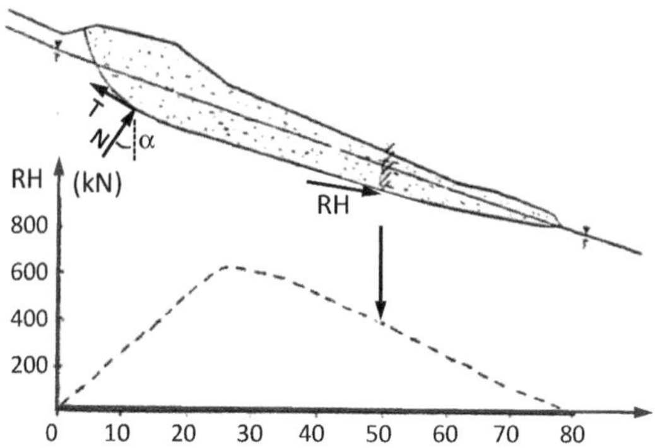

Figure 7.21. *Evaluation of the forces generated by the sliding mass at the top of the slip (Cartier 1985)*

7.7.2. Design and construction

Some time after the construction of a new highway at Lixing in 1977, a landslide occurred in the marl and limestone slopes, angled at around 30° in the direction of the road (Figure 7.22). A stability calculation using the inverse method showed that the residual angle of friction on the sliding surface was around 20° (taking into consideration a cohesion c′ = 0 kPa).

The solution chosen reduced the quantity of fill required by separating the active mass into two parts (Delmas and Matichard 1986) (Figure 7.22):

– an upper block that stabilized the upper portion of the sliding mass, above the sliding surface (1);

– a lower block that helped in stabilizing the foundation of the upstream block along sliding surfaces (2) and (3).

It should be noted that the surface was protected against UV-radiation using panels of lightweight concrete whose displacement was compatible with the expected deformation of the geotextile wraparound facing (Figure 7.23).

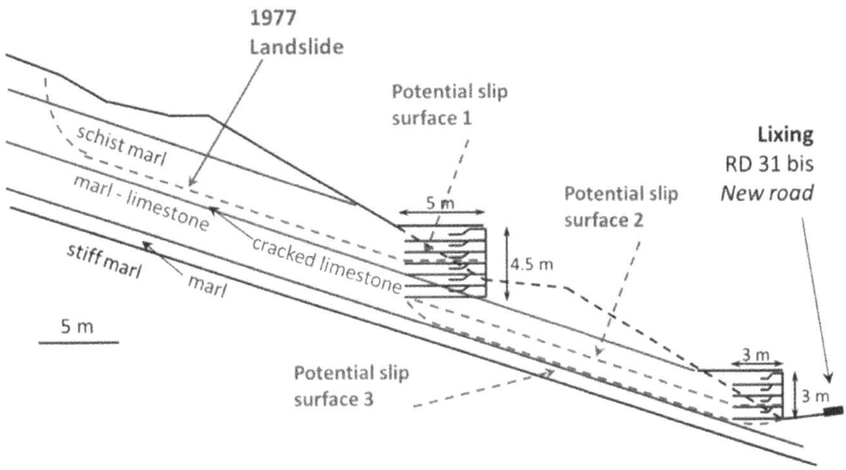

Figure 7.22. *Typical cross-section of the support solution for the Lixing landslide (1984)*

Table 7.7 presents the main geometric and geosynthetic characteristics of the two reinforced structures. The calculation (Figure 7.22) of the upper horizontal force (RH = 90 kN/m) permits an analysis of the stability of the upper block. As is usual in landslide stabilization calculations, an increase of 20% in the safety coefficient FoS was considered sufficient along the existing slip surface, and an FoS of 1.5 was considered for the other potential sliding surfaces.

Figure 7.23. *Photo of the structure supporting the Lixing landslide (1984). For a color version of this figure, see www.iste.co.uk/villard/geosynthetics.zip*

Lixing	1984	
Geometry		
Parameter	Block 1 (upper)	Block 2 (lower)
Height	4.5 m	3 m
Length	80 m	80 m
Angle of inclination of the facing	90°	90°
Geosynthetic		
Tensile strength	100 kN/m (1)	100 kN/m (1)
Stiffness	500 kN/m	500 kN/m
Spacing	0.75 m	0.75 m
Length	5 m	3 m
Soils		
Natural soil (marl and limestone)		
Angle of friction	$\varphi''_{peak} = 20°$ $\varphi''_{residual} = 20°$	
Cohesion	$c'_{peak} = 10$ kPa $c'_{residual} = 0$ kPa	
Unit weight	$\gamma = 19$ kN/m^3	
Reinforced backfill soil		
Angle of friction	$\varphi' = 35°$	
Cohesion	$c' = 0$ kPa	
Unit weight	$\gamma_d = 17$ kN/m^3	
Soil-geotextile interface friction	$\varphi'_{interface} = 25°$	

(1) Assumed value

Table 7.7. *Geometric characteristics and characteristics of the geotextiles in the reinforced blocks at Lixing (1984)*

The external stability of the two reinforced structures was ensured as follows:

– For the upper block, the conventional calculation of stability against overturning, sliding and bearing failure was carried out by considering the upper horizontal force (RH). From this, it was possible to derive the force that this block exerts on the foundation soil. The case presented here did not take into account any load uptake of the geosynthetic, except the interface shear force (no geotextile resistance for the facing).

– The forces obtained are then used to calculate the stability of the lower block. In this case, an FoS of 1.5 is required as the sliding surfaces under consideration are located in a region of the virgin soil mass.

The internal stability of the two blocks was ensured in accordance with rules in effect at the time using the "displacement method" (Delmas et al. 1986) and the critical internal slip surface (Figure 7.22). The tensile strength values for the geotextile and its stiffness are given in Table 7.7.

7.7.3. *Ageing and lessons learnt from the application*

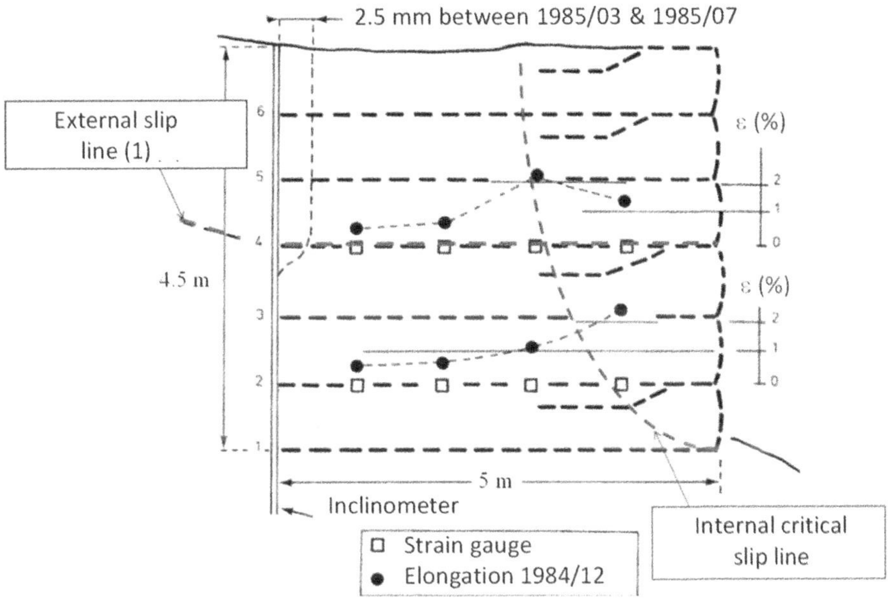

Figure 7.24. *Lixing (1984): deformation of the reinforced structure and extension strain in the geotextiles*

The instrumentation and monitoring of the reinforced blocks (especially the upper block) for a few months after its construction revealed the following behavior in the reinforced soil mass (Figure 7.24):

– Shortly after construction, the inclinometer installed immediately above the upper block registered a slight displacement (2.5 mm) in the section above the existing slip surface, a displacement that subsequently stabilized itself when the shear resistance was progressively mobilized. The relatively small angle of friction along the interface between the geotextile and the backfill justifies the width of the structure (5 m for a height of 4.5 m).

– At the same time, the deformations measured on the geotextiles seem to confirm a conventional mobilization of tensile forces that can be correlated to the internal stability of the reinforced block, but not to the upper horizontal force (RH) induced by the landslide. The external stability of the sliding zone is dependent on the mass of the reinforced soil blocks, but not on the mechanical characteristics of the geotextiles.

The satisfactory external appearance of the upper block, observed during a visit in 2016, confirmed the effectiveness of the geotextile solution. The only "damage" observed involved some eye-catching street art on the surface (Figure 7.25).

Figure 7.25. *View of the upstream block in 2016 (source: Photo V. Heili; Cerema Nancy 2016). For a color version of this figure, see www.iste.co.uk/villard/geosynthetics.zip*

7.8. Brides-les-Bains: reinforced segmental walls

7.8.1. *Concept and geotechnology*

From the mid-1980s onwards, segmental walls reinforced by geosynthetic sheets have been widely used in France because of their ease of construction and the aesthetically pleasing appearance of the facing (Gourc and Matichard 1994).

The specific case of Brides-les-Bains (Figure 7.26) will be considered here.

At this time, the cellular elements used were mainly Loeffel elements otherwise used as slope facings with an average inclination of 70°–80°. Without geosynthetic reinforcement at the rear of the structure, these segmental walls, used in soil embankments with standard mechanical properties, were said to be stable for slope heights of up to 3.5 m.

a) b)

Figure 7.26. *Brides-les-Bains: (a) reinforced segmental wall; (b) aesthetic appearance of the facing. For a color version of this figure, see www.iste.co.uk/villard/geosynthetics.zip*

a) b)

Figure 7.27. *Brides-les-Bains: (a) cellular facing of a geotextile-reinforced embankment (clamped between concrete cells); (b) cracks in the facing at the foot of the structure*

Geosynthetic inclusions clamped between two layers of cellular structures (Figure 7.27) enabled the construction of much higher reinforced embankments with a cellular facing.

7.8.2. *Design and construction*

The concept of combining a cellular concrete block facing with the geosynthetic-reinforced embankment was effective. However, the design calculations for the segmental wall had to be added to the design of the reinforced embankment, which by this time had been proven.

The Mezzo-Loeffel cells (effective height 0.26 m) were made of unreinforced concrete. For heights greater than the usual 3.5 m, cracks appeared where one cell unit rested on another (Figure 7.27(b)).

This type of reinforced cellular wall was used particularly for structure-supporting abutments (Figure 7.28(a)). They do not however support the structure deck, a task fulfilled by piles that are 1 m in diameter (Figure 7.28(b)).

a) b)

Figure 7.28. *Highway A26–(Chalons-Troyes): (a) reinforced embankment with cellular facing; (b) reinforced embankment with cellular facing and piles bearing the load of the bridge deck. For a color version of this figure, see www.iste.co.uk/villard/geosynthetics.zip*

7.8.3. *Lesson learnt from the application*

Another incident, involving the stability of the facing, made it possible to directly address the problem of designing the cellular facing. The structure was located on the A26 highway between Châlons-en-Champagne and Troyes, constructed in 1990. Figure 7.29 depicts an elevation view and a cross-section of the reinforced embankment. Unlike many earlier structures of this kind, for aesthetic purposes the designer had planned to have a cellular facing foundation that was not entirely horizontal but which rose on either side.

The embankment soil (c = 0, ϕ = 40°) was reinforced by woven polypropylene geotextiles (to eliminate the risk of chemical degradation by concrete) with the following characteristics: tensile rupture strength of 220 kN/m, with stiffness 2,000 kN/m. The design was carried out using the "displacement method", with the facing being considered as a soil (c = 50 kPa, ϕ = 40°).

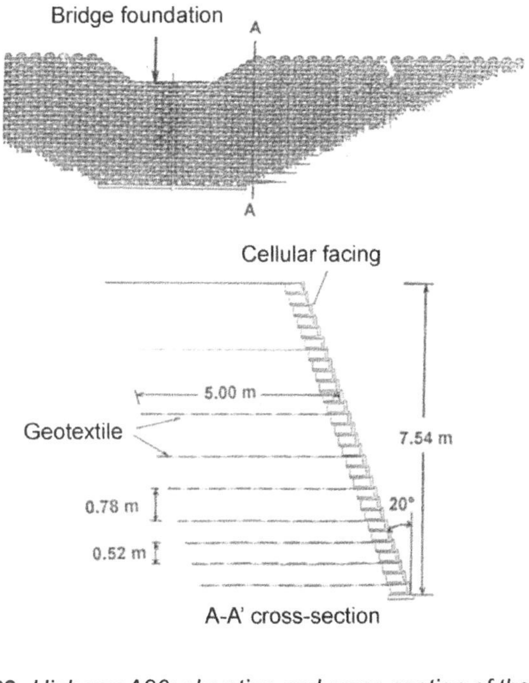

Figure 7.29. *Highway A26: elevation and cross-section of the structure*

The cellular facing had a slope of 70°. The spacing between the woven reinforcement layers varied, with an unreinforced height of 3.15 m (Figure 7.29).

With a uniform surcharge of 10 kPa and a safety coefficient of 1.5, the tensile force mobilized in the geotextiles did not exceed 40 kN/m.

The structure was completed in December 1990 and the failure occurred a few months later, but only in the right wing of the facing (Figure 7.30(a)). It is instructive to study the failure mode: the cellular wall slipped over its entire height (height between 2 and 5 m), with the slip axis situated 6.75 m from the structure's axis of symmetry.

It is remarkable that the reinforced soil behind the facing was not disturbed by the sliding of the facing. The reinforcing geotextiles, initially clamped between the concrete cells, were naturally freed from the cells. The excavation that was carried out after the incident clearly showed that the geotextiles continued to reinforce the embankment even in the absence of the cellular facing (Figure 7.30(b)).

Figure 7.30. *A26: (a and b) Failure observed in 1991, shortly after construction; (c) post-failure excavation with the reinforced embankment staying in place; (d) the left wing of the cellular wall remained stable. For a color version of this figure, see www.iste.co.uk/villard/geosynthetics.zip*

It can also be seen that there was no failure in the central section, where the wall is higher, but only on one of the wings of the wall. The reason for this was the sloping foundation of the lateral parts of the segmental wall, with erosion at the foot of the lowest cell destabilizing the base.

This incident led to greater vigilance in studying the stability of reinforced embankments with cellular facing with stability studies taking into account the study of the reinforced embankment, that of the facing and, finally, the combination of the two.

With respect to the components of the concrete cellular wall, new cells were designed specifically for the greater loads resulting from higher embankments, and since then this technique has become widely used around the world.

A visit to the Brides-les-Bains site in November 2017 (Figure 7.31) showed that despite the cracks in the facing, the structure did not exhibit any stability issues.

a) b)

Figure 7.31. *Brides-les-Bains (November 2017): (a) view of the whole structure; (b) view of the cells and a geotextile sheet. For a color version of this figure, see www.iste.co.uk/villard/geosynthetics.zip*

7.9. Gif-sur-Yvette (1988): retaining structure

7.9.1. *Concept and geotechnology*

The use of fiber-reinforced soils is a relatively old technique. However, Leflaive (1982, 1988) proposed the innovative process of using continuous filaments (polyester fibers) randomly distributed within a sand mass (Figure 7.32(b)). This solution, called Texsol, is today marketed by Eiffage. In the earliest applications of this micro-reinforcement system in France, these were 167 or 330 dtex polyester fibers (diameter 125 μm and 175 μm, respectively). The fibers are simultaneously distributed in the soil, along an inclined plane at an angle of 10–15°. This process requires a special machine whose size depends on the ground conditions. Figure 7.32(a) shows an example of this.

a) b)

Figure 7.32. *(a) Plant for the on-site production of micro-reinforcement: the continuous filaments blend with the sand; (b) view of the sand-fiber composite. For a color version of this figure, see www.iste.co.uk/villard/geosynthetics.zip*

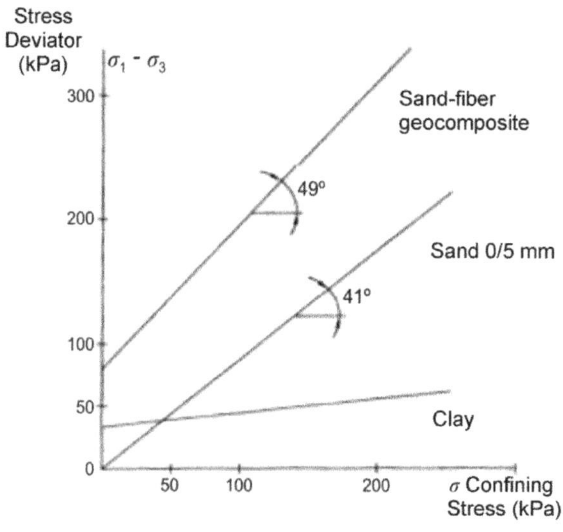

Figure 7.33. *Triaxial tests on the sand-fiber geocomposite: the shear strength of the soil-fiber geocomposite gives a cohesion of the order of 80 kPa for 0.1% PET fibers by mass. The micro-reinforced soil has a slightly greater angle of friction t*

Triaxial tests carried out on the reinforced material showed that the mechanical strength and the stiffness of the composite increase with the fiber content up to an asymptotic upper limit. In the standard case, the fiber content is 0.2% by weight, the D_{max} of the sand is 10 mm and D_{50} is equal to 0.50 mm (Khay et al. 1990).

The angle of internal friction of the composite is always greater than the angle of friction of the nonreinforced soil and an increase in the percentage by weight of the fibers results in an increase in apparent cohesion. The mean values obtained from the triaxial tests (Figure 7.33) are: $\phi' = 49°$ and c = 80 kPa for a percentage by weight of 0.1% of PET fibers with respect to the weight of the sand, for 167 dtex fibers and $\gamma_d = 18$ kN/m^3 for the sand.

7.9.2. *Design and construction*

Considering the improvement in the soil characteristics thanks to micro-reinforcement, one of the earliest applications of fiber-reinforced soil was in the construction of retaining walls (Figure 7.34) with a steep facing (between 65° and 75°) as in the example of Gif-sur-Yvette (Figure 7.35). The in situ soil behind the structure was a highly unstable sandy soil. The maximum height of the retaining wall was 7 m and the volume of micro-reinforced soil 600 m^3.

Figure 7.34. *View of a retaining wall built using micro-reinforced soil behind a segmental wall. For a color version of this figure, see www.iste.co.uk/villard/geosynthetics.zip*

a) b)

Figure 7.35. *Gif-sur-Yvette in 1988: (a) view of the micro-reinforced structure; (b) typical profile of the micro-reinforced wall. For a color version of this figure, see www.iste.co.uk/villard/geosynthetics.zip*

7.9.3. *Ageing and lessons learnt from the application*

Figure 7.36 shows a view of the structure in 2016 which can be compared to the view at the end of construction in 1988 (Figure 7.35(a)). One of the key questions related to the application of this technique is the vegetating of the facing, since the water retention in the reinforced soil is rather low. Figure 7.36(b) depicts the vegetation on the facing in 2016.

Lavaud et al. (2011) have presented interesting data on the ageing of the fibers. These data are obtained from an experimental micro-reinforced embankment constructed in 1987. Twenty years later, in 2008, fibers gathered from the surface of the embankment were subjected to mechanical and chemical tests. The sand is a 0/5 mm and the fibers were 167 dtex. A loss of molecular mass (15%–25%) was observed on the south and west facing, in contrast to the facing to the north (0% loss). On the other hand, a large and equivalent loss in tensile strength was observed on the north and south facings. The mechanical degradation cannot therefore be correlated with the chemical degradation. This does not agree with the findings from Risseeuw and Schmidt (1990), where a correlation was observed between the loss of molecular mass and the drop in tensile strength. According to these authors, the reduction in mechanical properties could be related to the compacting process or the sampling process.

a) b)

Figure 7.36. *Gif-sur-Yvette in 2016: (a) view of the micro-reinforced retaining structure; (b) close-up of the facing and the fibers on the surface. For a color version of this figure, see www.iste.co.uk/villard/geosynthetics.zip*

7.10. Frontenex (1992): cover for a spherical gas tank

7.10.1. *Concept and geotechnology*

Environmental concerns and the need to protect communities against major technological risks have resulted in significant changes to regulations governing the storing of substances such as liquefied gas. The regulations, which were reinforced by the Seveso directives in 1982, apply to existing facilities as well as new constructions. This has forced designers and contractors to look for new types of structures (Figure 7.37).

In this environmental context, the kind of micro-reinforcement presented in the preceding paragraphs may be a useful technique. It has been shown that a 60-cm layer of micro-reinforced sand offers protection equal to a 1 m thick layer of soil with respect to resistance against heat, explosions, military missiles, etc.

A polyester fiber retains 70% of its strength after 3 days at 175°C. With respect to resistance to explosions, experiments carried out using micro-reinforced soil have shown that projectiles such as shells or missiles, which created an 80-cm hole in a vegetated soil mass, penetrated to a depth of only 25 cm in a micro-reinforced layer.

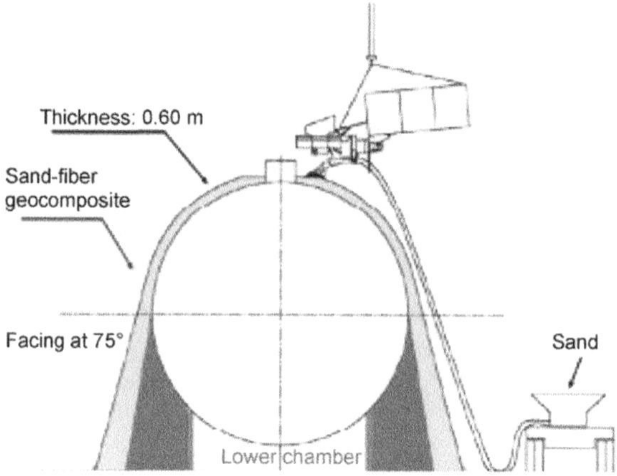

Figure 7.37. *Typical cross-section of a micro-reinforced covering layer on a spherical gas tank during construction. For a color version of this figure, see www.iste.co.uk/villard/geosynthetics.zip*

7.10.2. *Design and construction*

a) b)

Figure 7.38. *Micro-reinforced covering on the spherical gas tank: (a) covering made of micro-reinforced sand on the spherical Frontenex tank, just after the construction of the mound in 1992; (b) view of two protected tanks, separated by a short distance. For a color version of this figure, see www.iste.co.uk/villard/geosynthetics.zip*

The Frontenex tank was the first application of this kind (Figure 7.38(a)). A mound of micro-reinforced sand was constructed around the gas tank. With respect to fire-resistance, a 50-cm thick micro-reinforced layer is guaranteed to resist fire for at least 5 h (Khay and Claquin 2000). One further advantage of this geotechnology is that the distance between adjoining tanks can be reduced (Figure 7.38(b)).

Figure 7.39 depicts the construction of the protective layer, and Figure 7.38(a) shows the mound after plant cover was established.

Figure 7.39. *View of the micro-reinforced sand covering during construction. For a color version of this figure, see www.iste.co.uk/villard/geosynthetics.zip*

7.10.3. Ageing and lessons learnt from the application

Figure 7.40(a) depicts highly satisfactory appearance of the mound, and Figure 7.40(b) shows a close-up of the surface of the micro-reinforced sand covering in 2016.

a) b)

Figure 7.40. *The Frontenex spherical tank in 2016: (a) micro-reinforced sand covering; (b) close-up of the surface. For a color version of this figure, see www.iste.co.uk/villard/geosynthetics.zip*

In 2016, after 25 years of service, a new French regulation on earthquake-resistance required the reinforcement of the foot of the tank. This led to the mound being dismantled (Figures 7.41(a) and (b)).

Since the technique had proven to be satisfactory, the mound was reconstructed using the same micro-reinforcement process.

a) b)

Figure 7.41. *Dismantling of the covering around the Frontenex tank in 2016: (a) dismantling of the micro-reinforced sand covering; (b) behavior of the micro-reinforced soil during the dismantling of the mound. For a color version of this figure, see www.iste.co.uk/villard/geosynthetics.zip*

The geotechnology using micro-reinforcement has great potential for further development, although so far this has been sadly under-exploited.

7.11. Trois Lucs à la Valentine (1990): reinforcement over cavity

7.11.1. *Design and geotechnology*

The construction of infrastructure over subsidence zones has always required heavy (and often expensive) reinforcement solutions, especially when the position of the potential cavities is not known (Figure 7.42). In cases where it is not possible to fill the cavities, possible solutions are the bridging of critical zones or a piled embankment. With the appearance of geosynthetics with high tensile strength, a new solution emerged to prevent a sudden collapse when the cavity reaches the surface. The geosynthetic anchored at the sides functions as a membrane above the cavity and supports the overlying structure, thus slowing down surface settlements (Figure 7.43).

Figure 7.42. *View of a localized collapse under the roadway at Trois Lucs à La Valentine in 1994. For a color version of this figure, see www.iste.co.uk/villard/geosynthetics.zip*

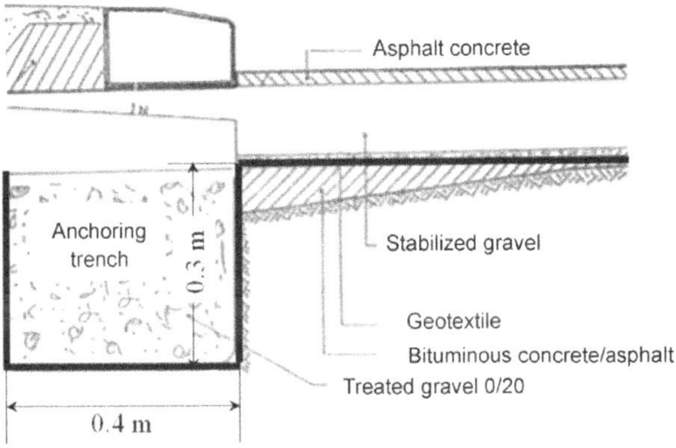

Figure 7.43. *Principle underlying geotextile reinforcement of a roadway over a potential localized collapse zone*

The mechanisms involved here were studied in the late 1980s and early 1990s (Gourc 1982; Perrier et al. 1990). The surface settlement is highly dependent on the ratio of the depth of the soil above the geosynthetic to the width of the cavity. This can be calibrated using the stiffness of the geosynthetic.

7.11.2. *Design and construction*

The Trois Lucs à La Valentine site is located close to Marseille in a suburban area. The geology of the site is marked by the presence of heterogeneous layers of marls and gypsum. The presence of underground galleries, constructed around 1922 at a depth of around 18 m, accelerated the decompression of the soils above these underground constructions. Several collapses have occurred in this zone, with problems accelerating toward the late 1980s (Figure 7.42).

A cement grouting system was proven to be to be ineffective.

The reinforcement of the roadway using a geotextile was the solution chosen and implemented at the end of 1990. Table 7.8 presents the main characteristics of the structure: geometry, geosynthetic materials, structure of the roadway and the anchoring of the geosynthetics.

Trois Lucs à la Valentine	1990
Geometry	
Road width	3.5 m
Surface diameter of cavity	2 m
Area to be secured	5,600 m^2
Anchorage length	1,600 m
Geosynthetic	
Type	GX 102 bimodular polyaramid polyester (1)
Tensile strength at 4% strain	120 kN/m (2)
Tensile strength at failure	200 kN/m (2)
Geotextile–geotextile seam strength	100 kN/m (2)
Chemical stability	4.5 < pH < 11.5
Thermal stability	Temperature < 110°C
Material for the roadway and anchorages	
Above the geotextile	Gravel 0/14 stabilized with bitumen Gravel 10/14 + cationic emulsion (18 kg/m^2)
Below the geotextile	Asphalt concrete 0/6
Lateral anchorages	Trench with treated gravel (depth 0.3 m, width 0.4 m)

(1) Product of Notex, current at the time.

(2) Manufacturer's data (the present-day tensile test standard was not published at that time).

Table 7.8. *Trois Lucs à la Valentine: characteristics of the geometry and the geosynthetics of the reinforced road structure over a zone likely to experience possible localized collapse*

The site owner had specified two main requirements for a cavity with a diameter of 2 m:

(A): a maximum settlement of 10 cm under an axle load of 35 kN;

(B): no failure of the geosynthetic under an axle load of 130 kN, but with no limit on the settlements in this case.

These requirements led to the development of an innovative, bi-modulus geotextile that blended polyamide fibers and polyester fibers using warp knitting technology.

The high stiffness of the geotextile up to 4% of deformation made it possible to fulfill requirement (A), while its resistance to rupture allowed it to fulfill requirement (B) (Figure 7.44(a)).

This technology also made it possible to obtain the force required for the liaison between two geotextile sheets (50% of the nominal resistance) through the use of polyamide strands (Figure 7.44(b)) between the loops at the edges of the geotextile sheets.

Figure 7.45 shows the installation of the geotextile in 1991. It was positioned between a layer of asphaltic concrete and a layer of gravel stabilized with bitumen. The lateral anchorage of the geotextile is also visible in this photo. The thermal and chemical stability of this type of geotextile (Table 7.8) reduced the risk of damage during installation and guaranteed its long-term performance.

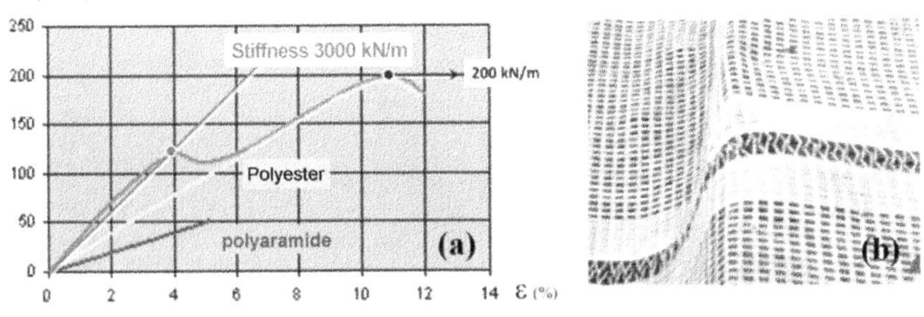

Figure 7.44. *Trois Lucs à la Valentine: (a) tensile behavior of the bimodular geotextile; (b) view of the seaming of two geotextile sheets. For a color version of this figure, see www.iste.co.uk/villard/geosynthetics.zip*

Figure 7.45. *Trois Lucs à la Valentine: view of the installation of the geotextile on the site in 1991. For a color version of this figure, see www.iste.co.uk/villard/geosynthetics.zip*

7.11.3. *Ageing and lessons learnt from the application*

As explained in the overall description of this project (Borie et al. 1991), in this type of application, tensile forces are not mobilized in the geotextile until a cavity reaches the surface. This happened toward the end of 2015 when a deflection of 8 cm appeared on the surface. This settlement resulted in the mobilization of the anchorage of the geotextile, as shown in Figure 7.46(b).

The geotextile solution was judged satisfactory by the Marseille roadways engineers, and they used the same solution to repair the road in 2016. Figure 7.47 shows the area repaired in 2016.

a) b)

Figure 7.46. *Trois Lucs à la Valentine: (a) pavement that underwent localized settlement in 2015; (b) view of the geotextile after the excavation of the collapsed zone in 2015. For a color version of this figure, see www.iste.co.uk/villard/geosynthetics.zip*

Figure 7.47. *Trois Lucs à la Valentine: area repaired in 2016. For a color version of this figure, see www.iste.co.uk/villard/geosynthetics.zip*

The structure of the reinforced roadway constructed in 1991 had proven to be effective in 2015. However, since the present-day design method is different, it is interesting to compare the 1991 method and present-day method for this case, taking into account the requirements enforced during the pre-installation study for a cavity with a diameter of 2 m (the dimensions of the cavity that was actually formed in 2015 are not known).

The present-day design method uses Eurocodes (the standard XP G38-065 (AFNOR 2020)) with a calculation of serviceability and ultimate limit states. In cases of reinforcement over potential cavities such as this, Eurocode Approach 2 is to be used.

In the context of an ultimate limit state calculation, the stability of the structure against the tensile failure of the geotextile, against the failure of the geotextile anchorage and against the excessive deformation of the structure must be verified.

The partial coefficients of the actions and the soils are given in Table 7.9. The reduction coefficients of the geotextile are given in Table 7.10.

Furthermore, in the context of the serviceability limit states, it must also be ensured that the maximum settlement value on the surface of the roadway for the entirety of its in-service life is not exceeded. The partial factors are all equal to 1, since they involve calculating the deformations and displacements under in-service conditions.

Eurocodes		Internal stability/block stability./ load-bearing capacity
Approach (Eurocodes)		Approach 2 A1 + M1 + R2
Actions		
Permanent		
Unfavorable	γ_G	1.35
Favorable	γ_G	1.00
Variable		
Unfavorable	γ_G	1.50
Favorable	γ_G	1.00
Soils		
Effective angle of friction	γ_{φ}'	1.00
Effective cohesion	γ_c'	1.00
Undrained cohesion	γ_{cu}	1.00
Unit weight	γ_{γ}	1.00
Geotextile		
Tensile strength	γ_{Mt}	1.25
Interface friction	γ_{Rf}	1.10

Table 7.9. *Partial factors taken into consideration in the Eurocodes – Approach 2*

The partial factors and reduction coefficients ($\gamma_{i;j}$ and Γ_i) corresponding to the ultimate limit state are applied to the geotextile's tensile strength at failure (for a deformation ε_{ult}) (Figure 7.48). The partial coefficients (C_i) corresponding to the serviceability limit state are applied for the deformation (ε) corresponding to the maximum expected settlement and are therefore applied to the geotextile's stiffness and not to its tensile strength (Figure 7.49).

	Reduction coefficients and partial factors		Geotextile on the site	1991 (perfect anchorage)	2017 (sliding anchorage)
SLS *Max. settlement 10 cm for an axial load of 35 kN*	C_{creep}	1	T at 4% = 120 kN/m	Max. diameter of the cavity 2.0 m	Max. diameter of the cavity 1.8 m
	C_{aeing}	1			
	$C_{installation}$	1			
	$\gamma_{M;t}$	1			
ULS *No rupture for an axial load of 130 kN* *No condition governing settlement*	Γ_{creep}	1	Tensile strength T_{ult} = 200 kN/m	Max. diameter of the cavity 2.0 m	Max. diameter of the cavity 1.9 m
	Γ_{ageing}	1.2			
	$\Gamma_{installation}$	1			
	$\gamma_{M;t}$	1.25			

Table 7.10. *Trois Lucs à la Valentine: comparison of the 1991 and 2017 design approaches – difference relating to the maximum permissible diameter of the cavity for the geotextile used*

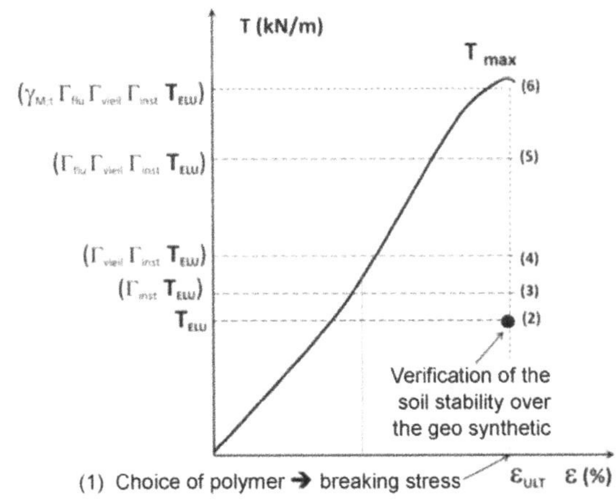

Figure 7.48. *Different steps in verifying the ultimate limit state corresponding to the tensile strength at failure of the geotextile*

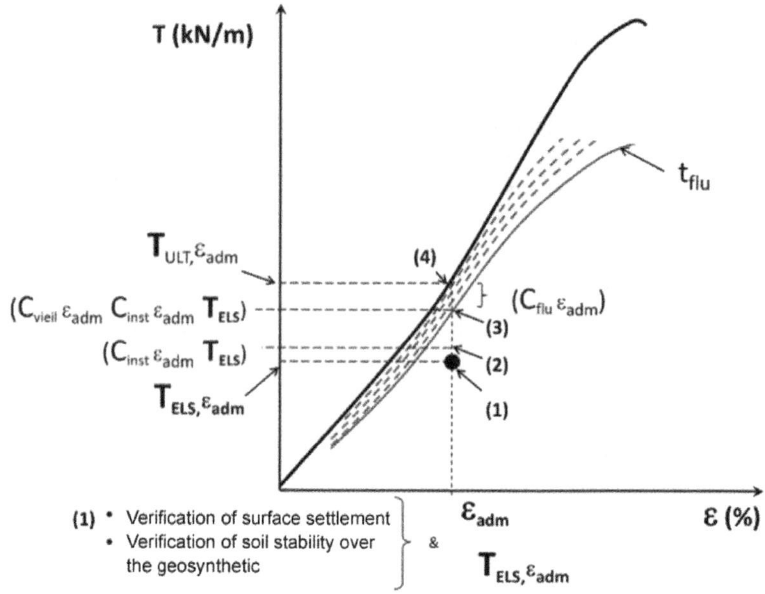

Figure 7.49. *The different steps in the verification of the serviceability limit state corresponding to the maximum surface settlement expected. The creep behavior of the product is directly accounted for in the isochronous curve for the in-service life. For a color version of this figure, see www.iste.co.uk/villard/geosynthetics.zip*

The other important difference between the design approaches of 1991 and 2017 is related to the lateral anchorage behavior. In the present-day approach, this is a sliding anchorage, while in the 1991 approach, it was rigid (perfect anchorage) (Delmas et al. 2015) (Figure 7.50).

It is evident that the 1991 design is less conservative than that of 2017:

– The 1991 calculation imposes a stiffness of 3,000 kN/m (T = 120 kN/m for $\varepsilon = 4\%$) instead of 4,600 kN/m (T = 184 kN/m for $\varepsilon = 4\%$), and a tensile strength at break of 230 kN/m instead of 210 kN/m.

– Table 7.10 shows that for the geotextile used here, the diameter of the cavity that can be stabilized according to the 1991 calculation is 10% greater (2 m instead of 1.8 m).

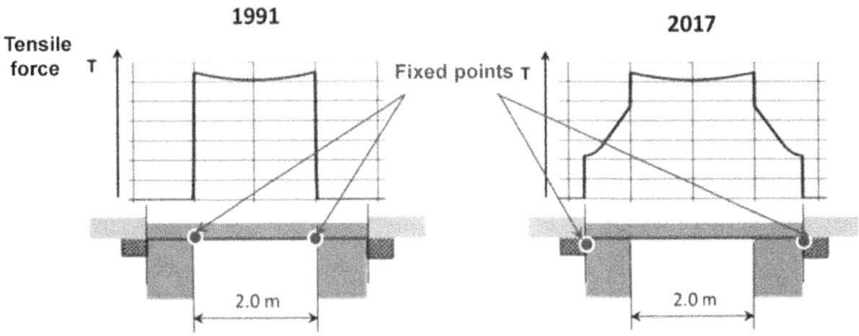

Figure 7.50. *Difference in the 1991 and 2017 design approaches. For a color version of this figure, see www.iste.co.uk/villard/geosynthetics.zip*

7.12. Roissard (1993): trench drain

Of all of the cases discussed here, this is the only structure that is still in service today but is not a permanent structure. This is an experimental structure located in Roissard, 30 km south of Grenoble. But trench drains instrumented and tracked over almost two decades are extremely rare, even at an international level.

7.12.1. *Concept and geotechnology*

Traditional drainage trenches are made up of clean gravel with a uniform grading. The gravel is separated from the in situ soil by a geotextile that acts as a filter (Figures 7.51 and 7.52).

In the site described here, it was required to drain a zone with a gentle slope within a glacio-lacustrine clay (varved clay). This is a critical filtration case, since the water can move along highly permeable strata (varves) or cracks, with the risk of transporting fine particles in suspension. The standard filter criteria for coarser, noncohesive soils are not adequate and we must consider the possibility of the filter becoming clogged.

In this context, a research program was launched in 1993 to study how relevant it would be to use a geotextile filter in these conditions (Figure 7.53).

7.12.2. *Design and construction*

The goal of this experimental study was to compare the effectiveness of various geosynthetic solutions. Six trench drains were dug, with depths of 2.0–2.5 m, a length of 12 m and a width of 0.60 m. Five of these trenches were filled with clean, round gravel (Faure et al. 1994, 1996).

Figure 7.51. *Typical cross-section of one of the trench drains. For a color version of this figure, see www.iste.co.uk/villard/geosynthetics.zip*

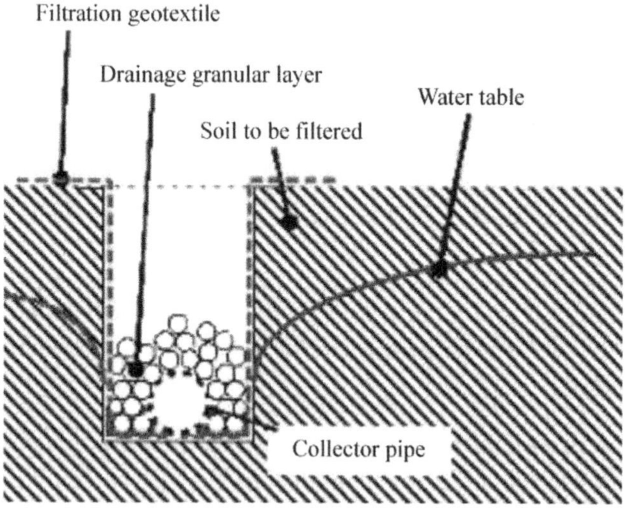

Figure 7.52. *Roissard (1993): view of a trench drain during construction, with the collector pipe at the bottom, the gravel and the geotextile filter. For a color version of this figure, see www.iste.co.uk/villard/geosynthetics.zip*

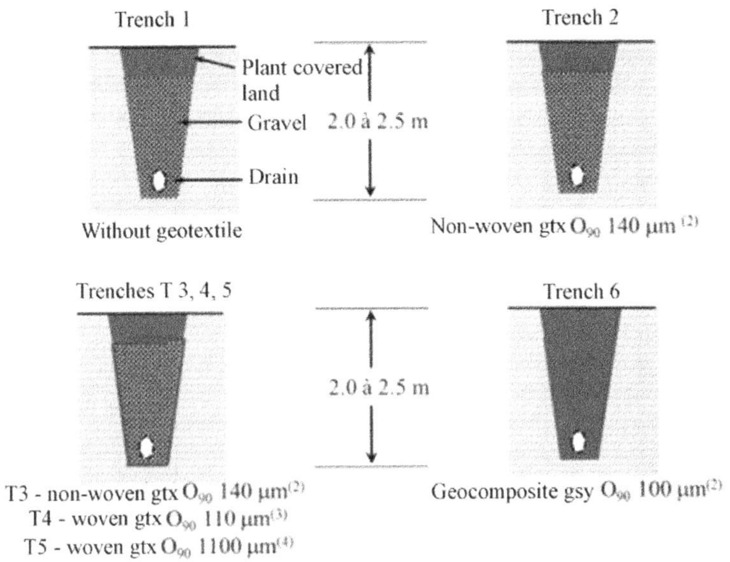

Figure 7.53. *Roissard: typical cross-sections of different trenches (geosynthetic depicted by the red line). For a color version of this figure, see www.iste.co.uk/villard/geosynthetics.zip*

– Trench T1, without geotextile, serves as the reference (Table 7.11 and Figure 7.53).

– Trench T2 has only a horizontal geotextile strip in the upper part of the trench.

– In trenches T3, T4, and T5, the gravel is wrapped within the geotextile.

– The final trench, T6, only has a geocomposite (no gravel) consisting of a thick geospacer wrapped in a geotextile filter.

The water outflow from each trench is collected separately. The suspended particles are also collected. Piezometers placed around the trenches make it possible to evaluate the drawdown in the surrounding water table.

The study of the outflows collected from the trenches (Table 7.12) shows that the flow from the reference trench, T1, without any geotextile is very small compared to the other trenches. Therefore, the outflow from trench T2 will be used as the reference outflow.

Roissard	T1	T2	T3	T4	T5	T6
Geometry						
Trench depth	2–2.5 m	2–2.5 m	2–2.5 m	2–2.5 m	2–2.5 m	2–2.5 m
Trench length	12 m	12 m	12 m	12 m	12 m	12 m
Geosynthetic						
Position	–	Top	Wrapped	Wrapped	Wrapped	Upstream side
Name	–	Bidim b2 (1)	Bidim b2 (1)	Propex 6,062 (2)	HaTe 43.144 (3)	Hydraway 2,000 (4)
Structure	–	Nonwoven	Nonwoven	Woven tape	Coated PET mesh	Drainage composite
Mass per unit area filter	–	130 g/m²	130 g/m²	190 g/m²	200 g/m²	233 g/m²
Drainage core thickness	–	–	–	–	–	3 cm
Filtration opening size (m)	–	140 (5)	140 (5)	110 (5)	1 100 (5)	100 (5)
Drain						
Diameter	160 mm					
Soil						
Soil type	Varved clays					
Uniformity coefficient	6					
< 2 μm	25%					
< 9 μm	85%					
< 20 μm	90%					

(1) Bidim geosynthetics at the time.

(2) Amoco.

(3) Huesker Synthetic GmbH.

(4) Hydraway.

(5) Filtration opening size measured using the hydrodynamic method (old French standard).

Table 7.11. *Roissard: characteristics of trench drains*

There is a sharp decrease in the T2 outflow over time when compared to the outflows from T3 (nonwoven filter) and T6 (geocomposite). For T4 and T5, wrapped in a geotextile with large filtration opening sizes, the outflow decreases but does not become critical.

	Mean outflow (L/min)		O_{90}/filter
Year	1993–1999	2008–2009	–
Trench T1	0.758	0.441	(No filter)
Trench T2	18.736	3.364	(single top filter)
Trench T3	5.856	5.905	140 µm/nonwoven
Trench T4	6.244	2.118	110 µm/woven
Trench T5	6.456	3.941	1,100 µm/woven
Trench T6	3.50	3.424	100 µm/geocomposite

Table 7.12. *Roissard: evolution of flow over time for the different trenches and different filtration opening sizes*

7.12.3. *Ageing and lessons learnt from the application*

After 18 years of tracking, it was decided in 2011 that all of the trenches would be excavated. It was thus possible to collect a large amount of data (Mériaux et al. 2013).

A few photos of the final state are given in Figure 7.54 for the reference Trench T2. Figure 7.55 depicts Trench T3 with the nonwoven filter. The other trenches are not presented in detail here. A more detailed analysis of the behavior can be found in Veylon et al. (2016).

The contamination of the gravel is significantly greater in Trench T2 when compared to T3. However, the collector pipe was not clogged in either case, even though there was partial blockage of the openings at the base of the T2 collector.

A thick calcite crust was observed on the clay/gravel interface in T2 and on the downstream face (gravel side) of the geotextile in T3. This corresponds to the regions where the porosity changes, which are undoubtedly correlated to the transition toward an aerobic zone.

In T3, caking with a thickness of 1–2 cm is observed on the upstream face of the geotextile (Figure 7.55).

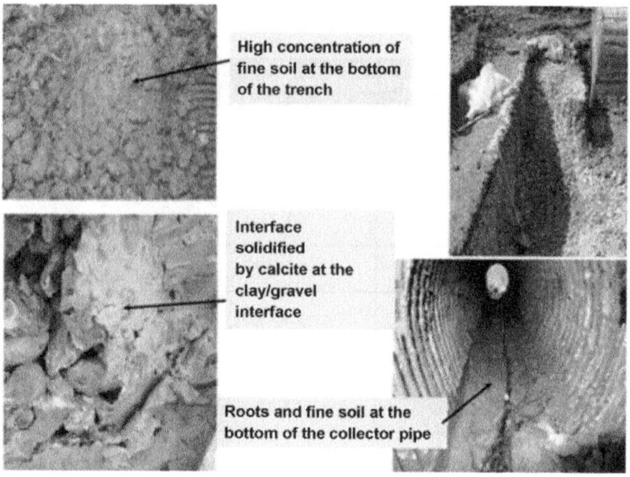

Figure 7.54. *Roissard: observations from 2011 for T2, without any geotextile. For a color version of this figure, see www.iste.co.uk/villard/geosynthetics.zip*

Figure 7.55. *Roissard: observations from 2011 for trench T3, with the nonwoven geotextile. For a color version of this figure, see www.iste.co.uk/villard/geosynthetics.zip*

Table 7.13 makes it possible to compare the behavior of different trenches by combining micro- and macro-observations. From this, it appears that the nonwoven geotextile is the best-performing solution for this site, but no general conclusions can be drawn from this.

It is interesting to compare the behaviors of trenches T3, T4 and T5 (Figure 7.53). All of these trenches have wrapped geotextile filters but have different filtration opening sizes and textures (Table 7.13).

Trench	Cake thickness	Calcite	Clogging (%)
T 2	–	+++	73
T 3	10–20 mm	+	21
T 4	20–40 mm	+++	16
T 5	2–10 mm	+	60

Table 7.13. *Roissard: observations from trenches T2 to T5 in 2011, during the excavation*

As specified in the introduction, this case study corresponds to the filtration of a cohesive soil, which is a case that has received little attention in literature. The behavior of geosynthetic filters is a complex phenomenon (Gourc 1982; Giroud 1985; Lafleur 1999; Cazzuffi et al. 2016; CFG 2016; AFNOR 2017).

A filter must essentially remain permeable to water, while also having the capacity for selective retention, maintaining the skeleton of the soil to be filtered while allowing the passage of finer particles to prevent clogging.

The retention capacity of a geosynthetic has been the subject of several proposed retention criteria. In particular, the recommendations by French IGS Chapter (CFG 2016), as well as the French standard G38-061 (AFNOR 2017), propose the following to allow for the finest fraction of soil to pass through the filter:

$$0_{90} \geq 63\ \mu m$$

The same recommendations from the CFG propose the following to ensure the retention of the soil skeleton:

$$0_{90} \leq C\ .\ d_{soil}$$

where $d_{soil} = d_{85}$ for a well-graded soil with wide granulometry, $d_{soil} = d_{50}$ for a soil with narrow granulometry:

$$d_{85} = 9\ \mu m \text{ and } C = C_1 \times C_2 \times C_3$$

Depending on:

– granulometry (continuous and wide $C_1 = 1$; uniform $C_1 = 0.8$) → $C_1 = 1.0$;

– soil density (loose $C_2 = 0.8$ – dense $C_2 = 1.25$) → $C_2 = 1$;

– hydraulic conditions (low hydraulic gradient and unidirectional flow) $C_3 = 1$; (high hydraulic gradient and two-way flow $C_3 = 0.6$) → $C_3 = 1.0$.

In the case presented here, $0_{90} \leq 9\ \mu m$.

This criterion is therefore not appropriate for fine, cohesive soils. The filtration opening size would be too small. For a plasticity index < 12, and if $0_{90} < 80\ \mu m$, the maximum value of 80 μm needs to be considered:

$$\rightarrow 63\ \mu m < O_{90} < 80\ \mu m$$

In the case presented here, the geotextile filters do not satisfy the retention criterion. The filtration opening size of all of the geotextiles is outside the validity range, even for T3 (O_{90} = 140 μm) which shows the most satisfactory long-term behavior.

This is not entirely surprising, given that the retention criteria are based on purely mechanical blockage by granular particles, without considering any adhesion between the particles of the present cohesive soil.

7.13. Valcros

7.13.1. *Design and geotechnology*

In 1970, a nonwoven geotextile of continuous polyester filaments was used in an earth dam (Figure 7.56) to fulfill major hydraulic functions. According to the authors, this was the first time a non-woven geotextile was used for this purpose.

This 17-m-high dam, designed by Giroud, is located in Valcros. This is a homogeneous dam of silty sand (Figure 7.57). The geotextile was used in two locations: the toe drain (mass/unit area 300 g/m^2) and the protection of the upstream facing (mass/unit area 400 and 850 g/m^2).

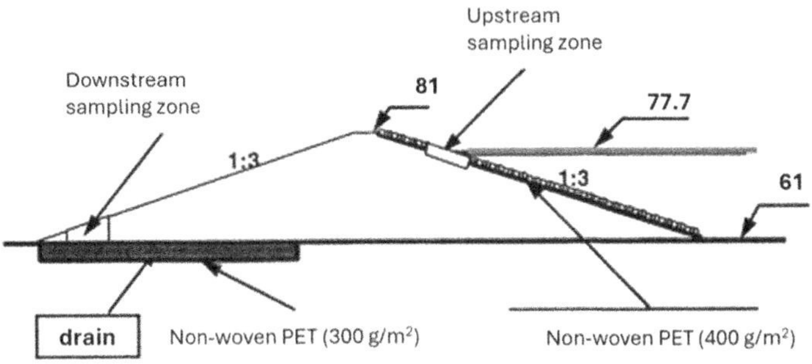

Figure 7.56. *Cross-section of the Valcros dam showing the location of the upstream and downstream geotextiles*

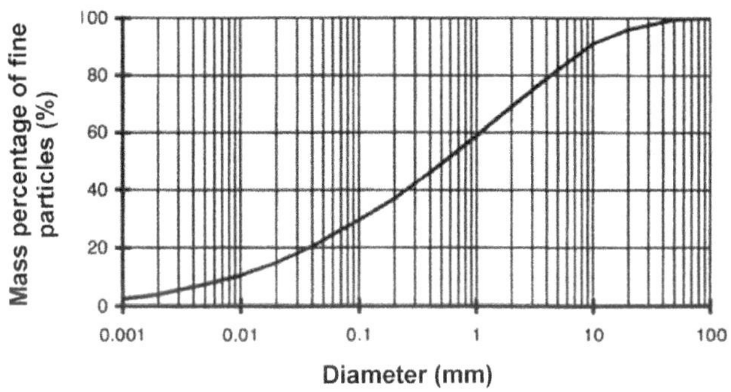

Figure 7.57. *Soil granulometry of the homogeneous earthen dam in Valcros*

7.13.2. *Design and construction*

7.13.2.1. *Upstream face*

At the time of construction, mineral materials were traditionally used as protection against wave action (Figure 7.58(a)). On an experimental basis, the upstream face of the Valcros dam was divided into three zones. The left-most zone had conventional protection with 250-mm blocks placed on a 150-mm-thick layer of 6/40 mm gravel; the central region had only a geotextile cover (mass/unit area 850 g/m^2) made of sheets sewn together and pinned in place on the soil, and the right-most area was protected by 250-mm blocks identical to those on the left-most zone, but placed on a layer of geotextile (mass/unit area 400 g/m^2) consisting of sheets with an overlap of 400 mm. When it failed in 1997, the central zone was reconstructed to match the right-most zone: rip-rap + geotextile.

7.13.2.2. *Toe drain*

The water percolating through the homogeneous earthen dam is collected by a gravel drain (Figure 7.58(b)). The innovative measure here was that it was a geotextile, instead of a traditional sand filter, that was wrapped around the gravel. This was a more economical choice as the sand that matched the filter criteria was not available locally and the geotextile was easier to install. This drain is made up of a collector fed by the water from the trench drains: 195 m^3 of 8/13 mm gravel wrapped in 1,450 m^2 of geotextile.

a) b)

Figure 7.58. *(a) Construction of the upstream face; (b) construction of the toe drain (1970). For a color version of this figure, see www.iste.co.uk/villard/geosynthetics.zip*

At the time of construction, in 1970, there were no filtration criteria for geotextiles, as it was not yet recognized that this material could fulfill this function. This novel application was thus implemented based on an empirical approach.

7.13.3. *Ageing and lessons learnt from the application*

Two sampling and measurement campaigns were carried out on the Valcros dam, one in 1976 (Giroud et al. 1977; Delmas et al. 1994) and it also featured in a study on the behavior of the geotextile (Faure et al. 1999).

During the last visit to the dam in 2014 (Figure 7.59), the dam showed strong long-term behavior.

a) b)

Figure 7.59. *(a) Downstream face; (b) upstream face (2014). For a color version of this figure, see www.iste.co.uk/villard/geosynthetics.zip*

7.13.3.1. *Upstream face*

The only incident to date on this dam was recorded in 1972 on the central zone of the upstream face, which was protected only by a geotextile. Because the geotextile sheet was not confined, the bank was eroded through wave action and gullies developed above the geotextiles (Figure 7.60(a)). This zone was repaired in 1972. The original geotextile was cut so that it matched the bank and a new protective measure, the same as on the right-most side (geotextile + rip-rap) was installed (Figure 7.60(b)). Since then, the combination has functioned well.

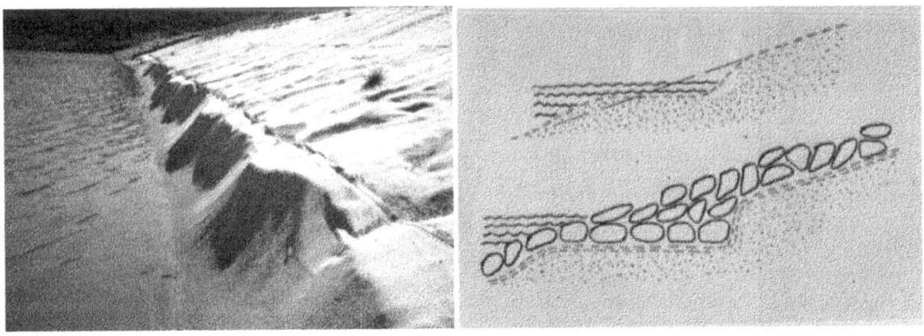

a) b)

Figure 7.60. *(a) Erosion in 1972 of the unconfined geotextile face; (b) erosion under the unconfined geotextile in 1972 and repair. For a color version of this figure, see www.iste.co.uk/villard/geosynthetics.zip*

Since then, we have learned through experience that it is essential to confine the geotextile to ensure the soil/filter contact and thus soil stability.

– The tests carried out on the samples taken from the wave-action zone on the upstream face, 6 years and 22 years after construction (Figure 7.61), showed that the percentage of geotextile voids filled with soil particles is greater between the rip-rap blocks (17%) than under the blocks (7%). The average porosity of the geotextile remained greater than 60%.

– The hydraulic conductivity of the sample, compared to an unused geotextile sample, is reduced by a factor of 10, without this affecting the performance.

Figure 7.61(b) shows a microscope photo of the geotextile between the rip-rap blocks (Faure et al. 1999).

Figure 7.61. *(a) Samples taken in 1992 from the upstream face; (b) view through a microscope of a cross-section of the sampled geotextile with soil on the upstream face*

7.13.3.2. *Downstream drain*

The performance of the drain has been satisfactory from the first time the upstream reservoir was filled:

– The water outflow from the drain is clear. Suspended particles were observed only for the first few days after the reservoir was filled.

– No seepage was observed on the downstream face.

– The flow rate at the drain outlet was in accordance with the estimate for an earth dam with hydraulic conductivity $K_s \approx 1 \times 10^{-7}$ m/s and a mean flow of 155 L/h.

It was, of course, not possible to take geotextile samples from the deepest part of the dam, corresponding to the hydraulic drawdown. An excavation was carried out in 1992 (Faure et al. 1999) over a length of 4.5 m, a width of 5 m and a depth of 4 m, so that the drain could be reached (Figure 7.62).

The tests carried out on the samples collected 6 years and 22 years after construction showed that:

– The hydraulic conductivity of the geotextile showed very little impact from being used as a filter, $K_f = 1 \times 10^{-3}$ m/s, a conductivity that only went up to 1.5×10^{-3} m/s after washing.

– Observation through a microscope confirmed that the clogging of the pores only occurred in 0.2% of the volume of interstitial pores.

Figure 7.62. *Excavation from 1992 and sampling from the foot of the dam at the drain outlet*

As mentioned above, at the time the dam was constructed there were no filtration criteria for geotextiles, and the criteria for granular filters could not be applied, especially because textile filters were much thinner.

Given that the drain behaved satisfactorily for over 50 years, it is interesting to compare present-day filter criteria with the soil characteristics and geotextile characteristics of the Valcros dam. Looking first at the filter criteria recommendations of the French Committee on Geosynthetics, and the French standard for the design of geosynthetics for draining a filter (NF G38-061; AFNOR 2017):

– Permeability criterion: measurements carried out by (Faure et al. 1999) give:

$k_s \approx 10^{-7}$ m/s $k_{g \approx}$ 1.5 × 10^{-3} m/s and geotextile thickness tg ≈ 3.5 mm

Since V_{H50}, corresponding to the current CFG standard, is not available, earlier recommendations based on the determination of permittivity $\psi_g = k_g/t_g = 1.15$ can be adopted:

$\psi_g > 10^{+5}.\ k_s$ verified, since 0.45 > 0.01

– Retention criterion: the filtration opening size of the geotextile is O_f = 100 μm. For the soil (Figure 7.63): $d_{85} = 6.4$ mm, $d_{50} = 400$ μm and the coefficient of uniformity Cu = 90.

Current CFG recommendations propose: $O_f < (C_1.C_2.C_3.C_4\ ).d_{50}$ (well-graded soil, wide granulometry), where:

– $C_1 = 1$ (wide granulometry);

– $C_2 = 1.25$ (dense, non-cohesive soil);

– $C_3 = 1$ (hydraulic gradient in the soil $i < 5$);

– $C_4 = 1$ (single filter).

Therefore, C = 1.25 and the retention criterion becomes $O_f < 500$ μm.

On the site, the geotextile opening size is $O_f = 100$ μm < 500 μm. This criterion is thus largely satisfied, but it leads to an oversized filtration opening due to the very wide soil granulometry (Cu = 90). Adhering to this criterion with an overly large opening, around 500 μm, could have led to soil leaching.

This is why Giroud (1985) suggested that only the fine fraction of soil (Cu = 3) be taken into account and that the coarse part of the soil is assumed not to form a skeleton of particles in contact with each other, but “floats” within the matrix of fine particles. Consequently, it does not contribute to the retention of finer particles.

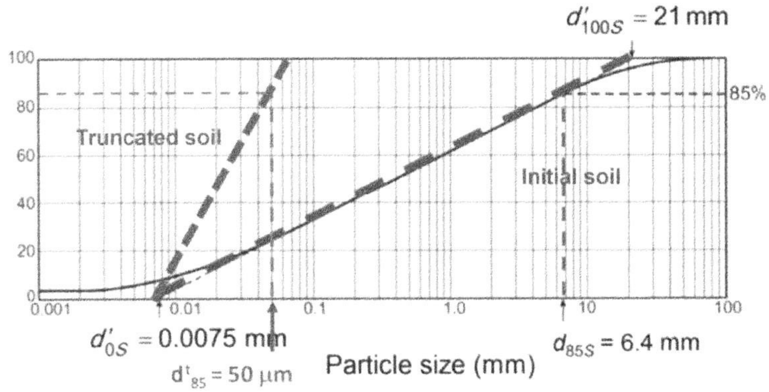

Figure 7.63. *Linearized granulometry curves for the initial soil found at Valcros and for the truncated soil (Cu = 3). For a color version of this figure, see www.iste.co.uk/villard/geosynthetics.zip*

Applying this principle leads to the use of the truncated soil granulometry (Cu = 3) where (d^t) is the diameter of the particles in the truncated soil, and (d_S) is the diameter of the initial soil (Cu = 90) (Figure 7.63).

The linearized granulometry curve proposed by Giroud (1985) is used. This is valid for soils with continuous granulometry, with the following formula:

$$(d^t_{100}/d^t_0)^{0.5} = (d^t_{60}/d^t_{10}) = 3 \text{ for the truncated soil (Cu)} = 3$$

with: $d^t_0 = d_{S0} = 0.0075$ mm (Figure 7.63), $d^t_{100} = 0.0675$ mm.

The truncated granulometry curve is plotted in Figure 7.63 and the CFG retention criterion is applied again but for the truncated soil.

The present-day CFG recommendations propose: $O_f < (C_1 \,.\, C_2 \,.\, C_3 \,.\, C_4).\ d^t_{85}$ (narrow granulometry), with:

– d^t_{85} = 50 µm (Figure 7.63);

– C_1 = 0.8 (narrow granulometry);

– C_2 = 1.25 (dense, noncohesive soil);

– C_3 = 1 (hydraulic gradient in the soil $i_s < 5$);

– C_4 = 1 (single filter).

Therefore, C=1 and the retention criterion becomes $O_f < C.d_{85} = 50$ µm, which is not satisfied since: $O_f = 100$ µm.

However, with respect to the CFG recommendations, if the retention criterion leads to $O_f < 80$ µm, it is enforced that $O_f < 63$ µm.

However, this criterion is still not satisfied, since $O_f = 100$ µm.

When applied to the initial soil, the criteria from literature cited by Faure et al. (1999) all lead to very large minimum opening sizes O. On the other hand, when applied to the truncated soil, they lead to values smaller than 100 µm.

However if we consider the retention criterion proposed by Giroud (1985) for dense soils:

$$O_f < 2.(Cu)^{0.3}.d'_{85} = 111 \text{ µm}$$

This retention criterion is satisfied since: $O_f = 100$ µm.

In conclusion, it is difficult to adapt retention criteria to soils with a very wide granulometry.

7.14. Ospedale (1979): upstream sealing on an earth dam

7.14.1. *Concept and geotechnology*

France has been a global pioneer in the use of geomembranes for upstream sealing of earth dams (Cazzuffi et al. 2010). The oldest example of this was the use of a butyl geomembrane on the Miel dam in 1968. From 1979 onwards, only two types of geomembranes have been used for this kind of application: a bituminous geomembrane, as used on the Ospedale dam, and a PVC geomembrane, as on the Aubrac dam (discussed in section 7.15).

Geomembranes are useful when there is no soil that can be used as a sealing layer. A geomembrane sealing system may be covered or exposed (Figures 7.64(a) and (b)). According to the ICOLD (2010) report, 30% of such structures have exposed sealing layers.

The Ospedale dam is one of the oldest tall dams (over 20 m high) with upstream sealing. As of 2019, the Ospedale dam was still operational (Figure 7.65).

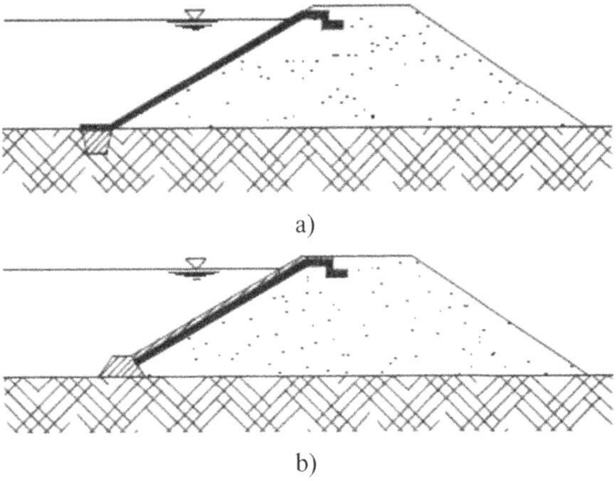

Figure 7.64. *Typical external upstream sealing system: (a) exposed geomembrane sealing layer; (b) covered geomembrane sealing layer*

Figure 7.65. *General view of the Ospedale dam in 2016. For a color version of this figure, see www.iste.co.uk/villard/geosynthetics.zip*

The Banegon dam (17m high, upstream slope of 2.5H:1V) was the first dam in France (1973) with an exposed bituminous geomembrane (Alonso et al. 1990). At this time, the geomembrane was made by the on-site impregnation of a nonwoven polyester.

In 1983, due to excessive leaks being observed downstream, the initial sealing was recovered by a bituminous geomembrane similar to the one used on the Ospedale dam. Since then the structure has behaved well (Gautier et al. 2002; Mafra et al. 2008).

The Ospedale dam, designed and constructed by SOMIVAC (The Society for Promoting Agriculture in Corsica (under the Bureau of Hydraulic Equipment in Corsica)), was 26 m high with a crest length of 135 m. The success of this project made it possible to construct the Ortolo dam in Corsica (1996). This dam was 38 m high and was built using the same technique.

7.14.2. *Design and construction*

The Ospedale dam is founded on a homogeneous, granite substrate. The body of the dam is made up of granitic rip-rap (0/1,000 mm). The slopes are 1.7H:1V upstream and 1.5H:1V downstream. The reservoir thus created has a water capacity of 3 million m^3 (Figure 7.66).

A cross-section of the sealing layer is depicted in Figure 7.67. A 4.8-mm-thick bituminous geomembrane (Colétanche NTP3) was installed between two nonwoven geotextiles with a mass/unit area of (400 g/m^2) (Figure 7.68).

The bituminous geomembrane (GBR-B) is a composite: a nonwoven, needle-punched polyester (300 g/ m^2) impregnated with bitumen (5,000 g/m^2) and reinforced by a fiberglass woven (50 g/m^2). From the 1980s onward, oxidized bitumen was replaced by an elastomeric bitumen, which has higher resistance to ageing and low temperatures.

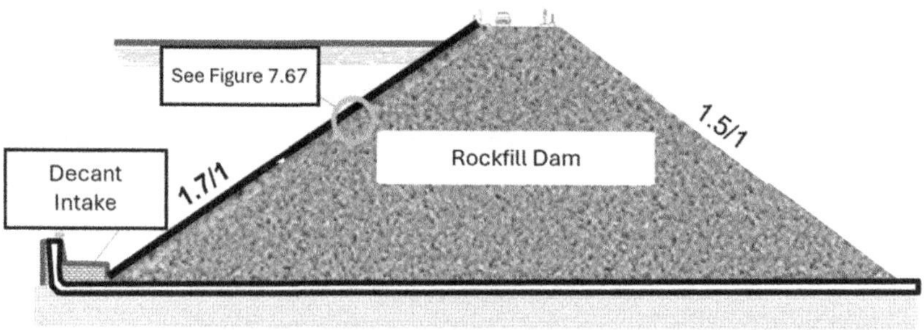

Figure 7.66. *Cross-section of the Ospedale dam. For a color version of this figure, see www.iste.co.uk/villard/geosynthetics.zip*

A protective layer of self-locking concrete blocks is arranged above it (0.08 m × 0.20 m × 0.12 m) with a unit mass of 8.5 kg.

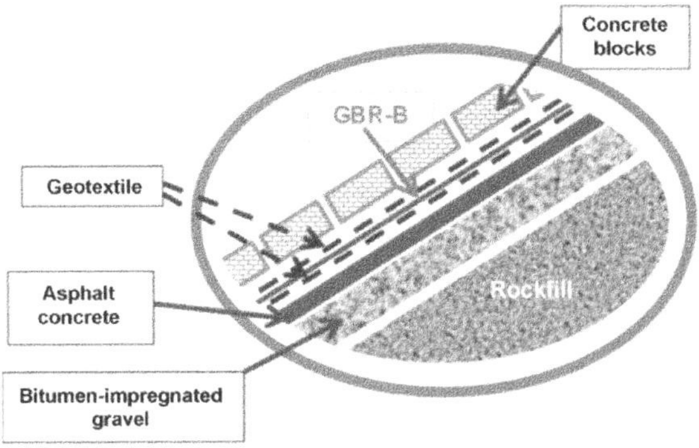

Figure 7.67. *Details of the upstream geomembrane sealing (GBR-B bituminous geomembrane). For a color version of this figure, see www.iste.co.uk/villard/geosynthetics.zip*

Figure 7.68. *Ospedale dam: installation of the sealing layer. For a color version of this figure, see www.iste.co.uk/villard/geosynthetics.zip*

The details of the jointing at the head and foot of the upstream face are given in Figure 7.69.

At the base of the geomembrane sealing system is a 2-mm-thick ballast layer (25/120 mm), a 100-mm-thick layer of gravel (25/50 mm) impregnated with a bituminous emulsion and, finally, a 50-mm-thick layer of bituminous concrete (3/6 mm), which can dissipate any eventual leaks.

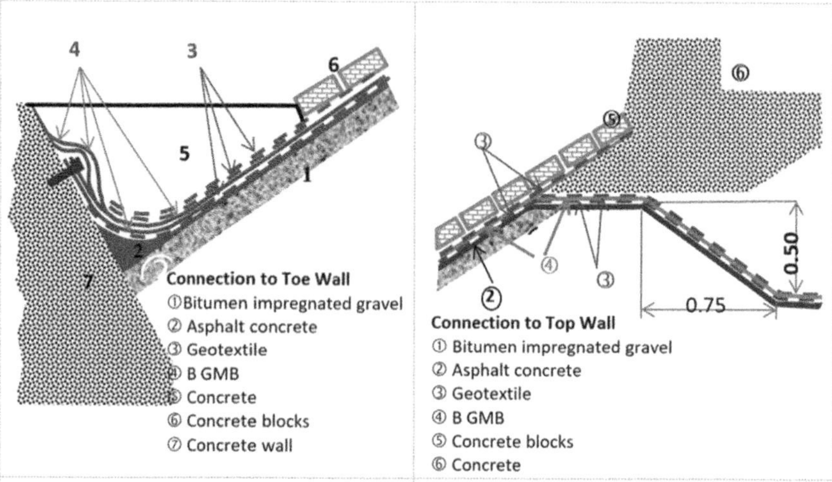

Connection of the GSS to the toe wall | Connection of the GSS to the top wall

Figure 7.69. *Ospedale dam: (a) connecting the sealing system at the foot; (b) connecting the sealing system at the crest. For a color version of this figure, see www.iste.co.uk/villard/geosynthetics.zip*

7.14.3. *Ageing and lessons learnt from the structure*

Figure 7.70 (Royet et al. 2002) depicts the evolution of the leakage outflow, collected in the drainage layer at the foot: just after construction, when the upstream reservoir was not yet full, the flow was practically double the flow observed during in-service conditions in the following two decades. The structure was completely drained in 1985, 1995 and 2006.

The average flow rate per unit area of the upstream face was 2.5 L/h/m^2, which is a moderate value. Cazzuffi et al. (2010) reports flow rates of between 1 and 10.0 L/h/m^2 for dams of the same type, and these leaks are attributed to the periphery rather than to the sealing system itself.

It is interesting to note that 2.4 L/h/m^2 $\approx 6 \times 10^{-2}$ m^3/m^2/day can be compared to the flow in a hydraulic conductivity test for a head height of 10 m, which is 10^{-6} m^3/m^2/day. This confirms that the leaks can be attributed either to the periphery or to the seams between sheets.

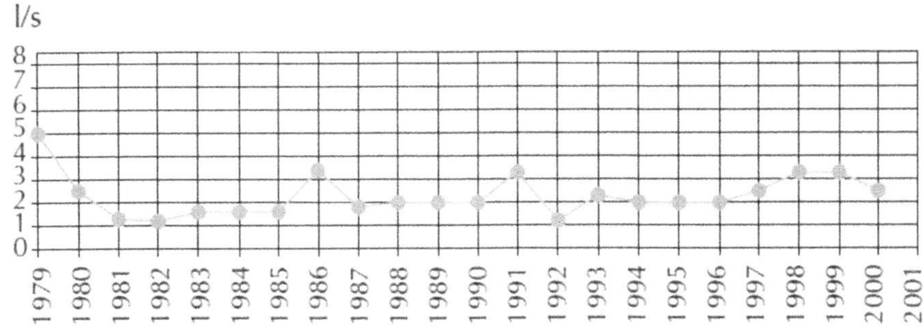

Figure 7.70. *Ospedale dam: leakage rate over time*

Figure 7.71. *Ospedale dam: self-locking protective blocks during the draining of the dam in 2005. For a color version of this figure, see www.iste.co.uk/villard/geosynthetics.zip*

A single, critical event for this structure deserves mention: in 1983, a severe storm created strong wave action, which resulted in a part of the layer of protective, self-locking blocks sliding down. Thanks to the protective upper geotextile, no damage to the geomembrane resulted from this incident. However, this showed that the surcharge created by these blocks was too light. The Ortolo dam (1996), which used the same sealing bituminous geomembrane, had 0.14-m-thick plates made of concrete reinforced with polypropylene fibers, with 20-mm joints to dissipate excess pressures. The disadvantage of this system, however was that it was not well-suited to carrying out repairs on the sealing layer.

In 2007, samples of the bituminous geomembrane were collected from three locations on the upstream face (Figure 7.72).

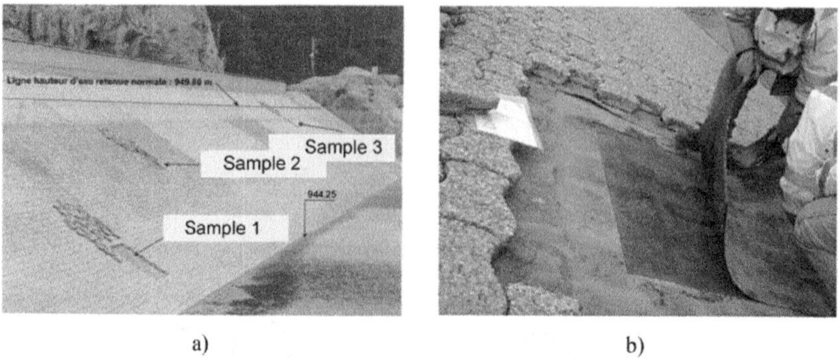

Figure 7.72. *Ospedale dam: sampling campaign: (a) location of the samples; (b) sampling of the GBR-B between two geotextile layers. For a color version of this figure, see www.iste.co.uk/villard/geosynthetics.zip*

7.14.3.1. *Hydraulic conductivity*

The flow across the geomembrane in a permanent regime was measured using an INREA cell corresponding to the standard NF EN 14150 (AFNOR 2019) (Figure 7.73) (Lambert and Touze-Foltz 2000; Touze-Foltz et al. 2010). The upstream pressure was 150 kPa and the downstream load was 50 kPa.

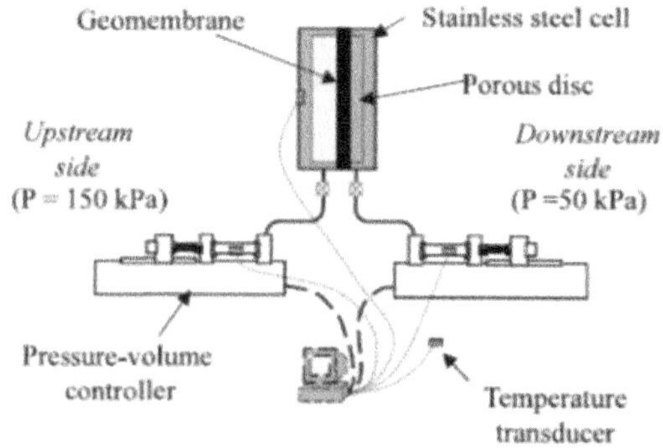

Figure 7.73. *Hydraulic conductivity test on the geomembrane*

The flow for all three samples taken was less than 10^{-6} m^3/m^2/day, which corresponds to the measurement limit of this equipment. Of interest is that this flow is 1,000 times lower than the flow through a 1-m-thick layer of clay with a hydraulic conductivity of 10^{-9} m/s.

7.14.3.2. *Mechanical characteristics*

Tensile strength tests were carried out in 2007 on three circular samples with a diameter of 0.20 m, on burst-testing equipment (NF EN 14151; AFNOR 2010). A "virgin" sample, supplied by the manufacturer, was tested for comparison (Table 7.14). The samples were subjected to hydraulic pressure, at 50 kPa/min, until water was observed to leak through the geomembrane.

Sample	Pressure (kPa)	Dome height (mm)	Deformation (%)	Ultimate tensile strength (Mpa)
"Virgin"	212	41	10	3.8
1	119	24	3	3
2	94	18	2	3
3	103	17	2	3.4

Table 7.14. *Results from the burst test on the bituminous geomembrane*

A spherical deformation is assumed and calculated based on the dome height. A decrease of 17.5% in the ultimate tensile strength was observed, while the stiffness doubled.

7.14.3.3. *Physical–chemical properties*

Sample	BRT (°C) EN 1427	Asphaltene content (%) NF T 60-115
"Virgin"	≈ 100 ± 5	22
1	139.5	29.3
2	144	22.7
Exposed GBR-B	150	35

Table 7.15. *Evolution of the physical–chemical properties of oxidized bitumen*

The corresponding tests were carried out at INREA in Aix-en-Provence. Two samples from the site were compared to two samples that had undergone extreme exposure, a "virgin" oxidized-bitumen sample (100/40) and an average sample corresponding to bituminous geomembranes that were exposed to UV radiation for 30 years at other sites. The tests focused on the softening point (ball and ring temperature (BRT)) and the asphaltene content (Table 7.15). An ageing of the bitumen can be seen but remains acceptable.

7.15. Aubrac (1986): upstream sealing of an earth dam

7.15.1. *Concept and geotechnology*

Since 1979, only two types of geomembranes have been used for this kind of application: a bituminous geomembrane, as used on the Ospedale dam (see section 7.14) and a PVC geomembrane, as used on the Aubrac dam (Alonso et al. 1990). The case of the Aubrac dam is interesting as the samples were collected on site.

7.15.2. *Design and construction*

The PVC geomembrane is placed on the dam's upstream face (Figure 7.74(a)) and covered with a protective basalt rip-rap 100/300 (Figure 7.74(b)). The geomembrane itself is placed between non-woven geotextiles. Figure 7.75 shows a detailed cross-section of the sealing layer, especially the design adopted to ensure continuity of the sealing at the toe of the structure, between the geomembrane and the basal grouting curtain: the geomembrane is anchored in the vertical sealing curtain (Figure 7.76).

a) b)

Figure 7.74. *Aubrac dam: (a) upstream view (2016); (b) installation of the protective rip-rap (1986). For a color version of this figure, see www.iste.co.uk/villard/geosynthetics.zip*

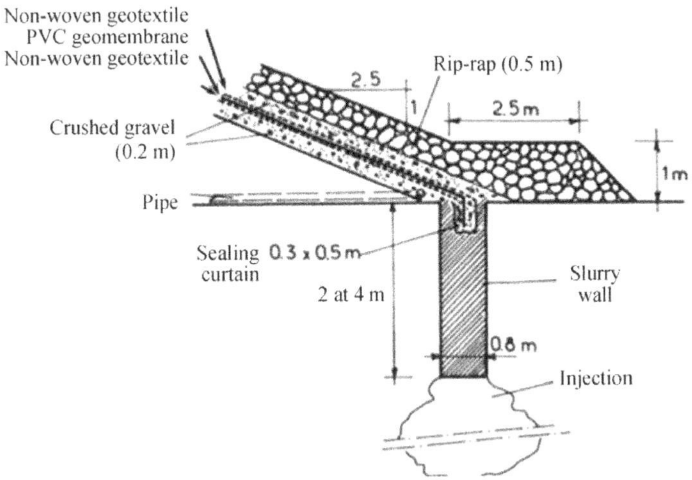

Figure 7.75. *Aubrac dam: cross-section of the upstream sealing at the toe of the dam*

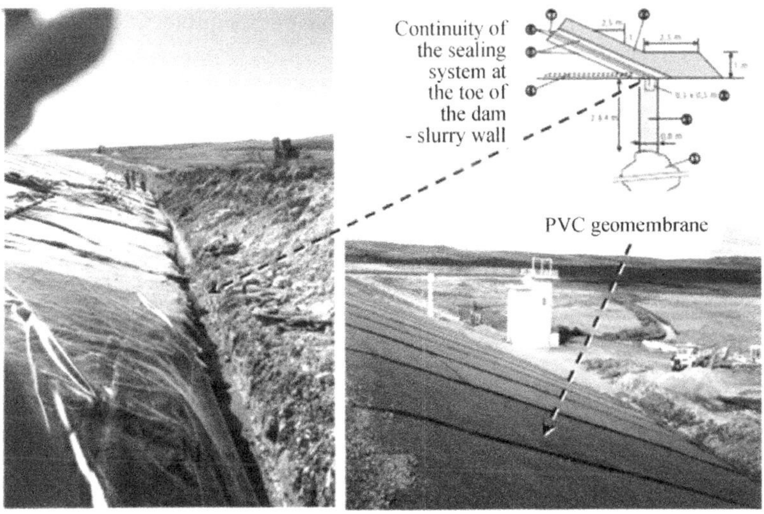

Figure 7.76. *Aubrac dam: connection between the PVC geomembrane and the vertical diaphragm wall to ensure continuity of sealing (1986). For a color version of this figure, see www.iste.co.uk/villard/geosynthetics.zip*

Table 7.16 presents the characteristics of the structure.

Aubrac		1986	
Geometry			
Height		14 m	
Length of the structure		356 m	
Angle of the upstream slope		22°	
Geosynthetics			
Type	Geomembrane Alkorplan PVC-P (1)	Upper protective nonwoven geotextile (2)	Lower protective nonwoven geotextile (2)
Mass/unit area	–	500 g/m²	500 g/m²
Thicknesses	1.2 mm	–	–

(1) Alkor-Maréchal product.

(2) Product manufactured by Vigano-Pavitex, product name unknown.

Table 7.16. *Aubrac dam: characteristics of the structure*

Figure 7.77 depicts the upstream face with its geomembrane sealing layer not yet covered. Figure 7.74(b) depicts the installation of the protective rip-rap. Figure 7.74(a) shows a recent image of the site.

Figure 7.77. *Aubrac dam: exposed geomembrane during construction (1986). For a color version of this figure, see www.iste.co.uk/villard/geosynthetics.zip*

Thirty years after construction, the structure still functions satisfactorily. The flow of seepage collected in the downstream drainage system is practically constant, around 3 L/s.

7.15.3. *Ageing and lessons learnt from the structure*

After an on-site sampling campaign (Royet et al. 2002) in 2000 (Figure 7.78(a)), tests were carried out on the PVC geomembrane (Figure 7.78(b)) that confirmed that the geomembrane had evolved in a satisfactory manner (Figure 7.78(c)):

– Burst tests showed that there was a slight increase in the tensile stiffness of the PVC geomembrane.

– The chemical analysis showed a slight loss in plasticizers, this loss being especially noticeable above the water level.

An incident occurred during the installation of the upstream protection (Figure 7.79). The laying of the gravel layer (an intermediate layer between the geosynthetics and the rip-rap) was modified by the gravel being emptied from the top of the upstream slope instead of starting from the base: the granular layer began to slide down the slope, dragging the upper geotextile with it as well as the geomembrane over about 10% of the upstream face (Figures 7.79 and 7.80).

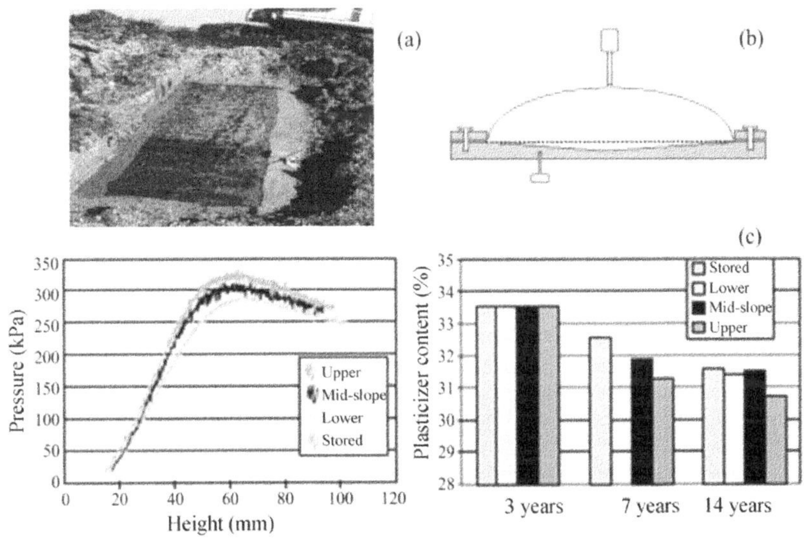

Figure 7.78. *Aubrac dam: (a) sampling campaign; (b) burst testing; (c) evolution of the bursting behavior over time and of the plasticizer content between construction in 1986 and collection of samples in 2000. For a color version of this figure, see www.iste.co.uk/villard/geosynthetics.zip*

a) b)

Figure 7.79. *Aubrac dam: view of the sliding of the geomembrane and the protective layers at the interface with the lower geotextile during construction (1986). For a color version of this figure, see www.iste.co.uk/villard/geosynthetics.zip*

Following this event, experiments were carried out with a new method to determine interface friction (inclined plane test), which made it possible to take into account a small stress normal to the interface (corresponding to the surcharge of the protective layer):

– The conventional test using a large shear box gives an angle of friction between the geomembrane and lower geotextile that is less than 34° (under the geomembrane).

– For the same interface, the use of a rough shear box test (Figure 7.80) gives an angle of friction of only 25° (Girard et al. 1990).

Figure 7.80. *Aubrac dam: detailed view of the sliding at the interface between the lower geotextile and geomembrane (1986). For a color version of this figure, see www.iste.co.uk/villard/geosynthetics.zip*

a) b)

Figure 7.81. *(a) The IRSTEA inclined place simulating the multilayer interface of the upstream face of the Aubrac dam; (b) sliding obtained by increasing the inclination. For a color version of this figure, see www.iste.co.uk/villard/geosynthetics.zip*

Since the slope of the dam's upstream facing is at an angle of 22°, the sliding that was observed could have been anticipated. The difference between 22° and 25° could be justified by the destabilizing, dynamic effect of the discharge of the gravel at the site.

Since then, the inclined plane test has been improved and has demonstrated how versatile it can be to use and its potential for simulating real-life scenarios (Carbone et al. 2015) under a small, normal stress.

7.16. Jonage (1994): bank protection

7.16.1. *Concept and geotechnology*

Geosynthetic concrete mattresses are an effective technique to restore the sealing of canals or reservoirs, with the new sealing lining covering the older one. One of the advantages of this technology is that it can be installed underwater, without having to reduce the water level.

A standard mattress consists of two woven polyethylene-polyamide geotextiles that are attached to each other. Liquid concrete is injected into the void between these two geotextiles on the site, with the geosynthetics acting as lost formwork. Today, we are more aware of the possible risk of plastic microparticles being released, although the risk is minimal in the present case. To remedy this problem, current recommendations may include using the same procedure with natural fiber formwork, like those used for erosion control.

This kind of a liner is quick to install, even underwater. The geotextile sheets are first cut to the correct size and factory assembled in large panels that are zipped together on site.

Pipes are used to fill the mattress with liquid concrete. The final mattress has a minimum thickness of 10 cm. Once installed, the sealing lining is implemented. The application discussed here was a pioneering installation for France. For new applications, additive UV protection by a vegetalized soil layer, or any appropriate technique, should be considered.

7.16.2. *Design and construction*

The Jonage canal in the Lyon region is a diversion of the river Rhone for hydroelectric reasons. Downstream of the Jonage dam, to the right of the canal, is a clay dike (Figure 7.82). This dike, whose upper part is covered in concrete, suffered from infiltration and seepage in a section close to the dam. Consequently, it was decided that a new sealing layer would be installed upstream of this dike over an embankment length of 356 m (Table 7.17). A typical cross-section is depicted in Figure 7.82 (Koffler 1995). The mattress was transported to the site in panels that were 36 m long and 33 m wide. A team of divers was brought to the site during the placing and the injection of the mattress (Figure 7.83(a)). It took three months to install the mattress.

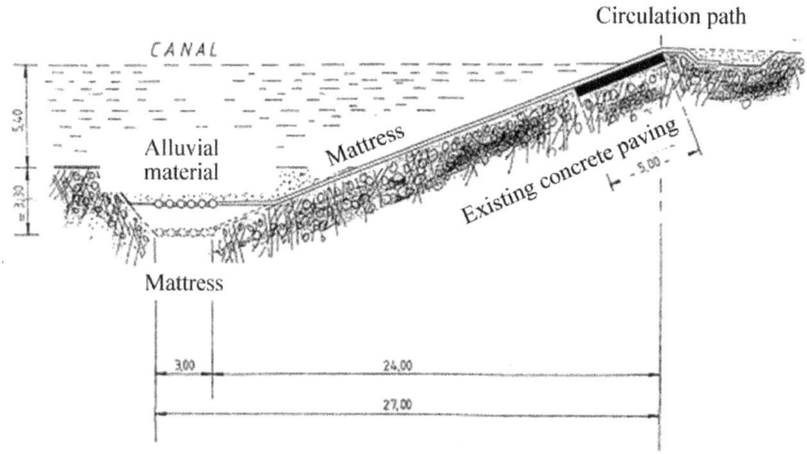

Figure 7.82. *Jonage: typical cross-section of the bank protection using a geosynthetic-concrete mattress*

Jonage	1994
Geometry	
Height	8.70 m
Length of the structure	356 m
Angle of the facing	20°
Geosynthetic	
Type	Incomat [(1)]
Polymer	PE/PA
Length	36 m

(1) Product made by Huesker.

Table 7.17. *Jonage canal: characteristics of the geosynthetic-concrete mattress*

To prevent erosion under the mattress, the first upstream panel was a flexible panel (Figure 7.83(b)) buried in a trench running crosswise to the flow.

a) b)

Figure 7.83. *Jonage: (a) installation of the geosynthetic mattress in 1994; (b) special flexible panel to prevent lateral erosion n under the standard mattress. For a color version of this figure, see www.iste.co.uk/villard/geosynthetics.zip*

7.16.3. *Ageing and lessons learnt from the application*

Figure 7.84(a) shows the end of the installation phase and Figures 7.84(b) and (c) are photographs of the site 22 years later. Although considered to be sacrificial shuttering, not useful after construction, the geotextile component is still present, except in zones subjected to wave action. In the zone that may be unprotected from UV radiation, the geotextile shall be protected with a soil layer in order to prevent degradation. In addition, there was no sign of any seepage on the downstream side of the dike in 2016 (Figure 7.84(b)).

Figure 7.84. *Jonage canal: (a) view of the sealing mattress at the end of the installation in 1994; (b) global view of the embankment downstream of the dike and the sealing mattress in 2016 (the dotted line shows the dike profile); (c) view of the geosynthetic-concrete mattress in 2016. For a color version of this figure, see www.iste.co.uk/villard/geosynthetics.zip*

7.17. Pont-de-Claix (1974): industrial reservoir

7.17.1. *Concept and geotechnology*

In June 1974, the first reservoir with a double geomembrane liner was constructed in Pont-de-Claix, near Grenoble. The reservoir belonged to a chemical factory (today the PERSTORP company).

This (emergency) reservoir is located at the top of a 50-m slope overlooking the chemical factory, at the foot of an upstream bank that is 90 m high. Its altitude is 245 m. A double liner was justified by the fact that a geotechnical study showed that there was a need to strictly control water leakage to prevent any instability in the slope (Figure 7.85), which could have resulted in pollution and release of gas from the factory.

The double liner design was first presented by Giroud (1973).

Double liner using geomembranes (Figure 7.86) involves two geomembranes separated by a drainage layer. The upper geomembrane is called the primary liner and the lower geomembrane the secondary liner. The function of the drainage layer is the detection, collection and removal of any possible leakage through the primary liner. An effective secondary–liner here can only be provided by a geomembrane because, as any leakage would normally be minimal, it would simply be absorbed by a clay layer, thereby preventing detection of the leak and the evacuation of the water from the site.

Figure 7.85. *Location of the reservoir of the Perstorp chemical factory in Pont-de-Claix. For a color version of this figure, see www.iste.co.uk/villard/geosynthetics.zip*

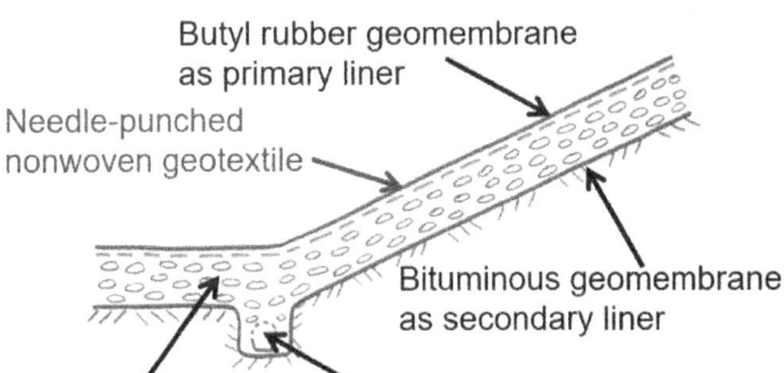

Figure 7.86. *Double liner used in the Pont-de-Claix reservoir. For a color version of this figure, see www.iste.co.uk/villard/geosynthetics.zip*

Any possible leakage had to be swiftly removed to ensure that the water level in the drainage layer remained low and, in all cases, less than the thickness of the layer. Thus, the hydraulic head on the secondary liner is low, and even one or two holes in this geomembrane will only result in minimal seepage into the underlying soil.

The flow capacity of this drainage layer is thus an essential factor: it must have an appropriate slope angle and adequate hydraulic conductivity.

The reservoir needed to be emptied within a few hours, if this were required by the chemical factory, but this has never been the case since its construction. Nonetheless, this meant that the primary geomembrane layer could not be covered by a granular layer, which could become unstable in the case of rapid drawdown. The shaded location of the reservoir made this choice possible.

7.17.2. *Design and construction*

The reservoir has a capacity of 40,000 m^3 and its slopes are 1V/2H. The water depth varies between 8 and 10 m and the sealing layer has an area of 10,000 m^2.

The reservoir cavity is almost totally in cut and thus did not pose any stability problems as long as there was complete protection against any risk of seepage.

The secondary liner is a bituminous geomembrane, manufactured in place by Colas, as was usually done until 1975 by pulverizing bitumen at 200°C (6–10 kg/m^2) on a nonwoven polyester geotextile (400–600 g/m^2). The melting point of polyester is 260°C. The final thickness of the secondary liner was between 8 and 10 mm (Figure 7.87).

a) b)

Figure 7.87. *(a) Pulverization of the bitumen for the secondary liner; (b) secondary layer after the pulverization of the bitumen. For a color version of this figure, see www.iste.co.uk/villard/geosynthetics.zip*

The 0.20-m-thick drainage layer (Figure 7.88), which is placed directly on the bituminous geomembrane, consists of round gravel (15–25 mm) stabilized by a small amount of mortar (100 kg/m^3 of sand, 100 kg/m^3 of cement added to 1,900 kg/m^3 of gravel). Due to the potentially aggressive nature of cemented sand, a 400 g/m^2 nonwoven polyester geotextile was placed between the drainage layer and the primary liner.

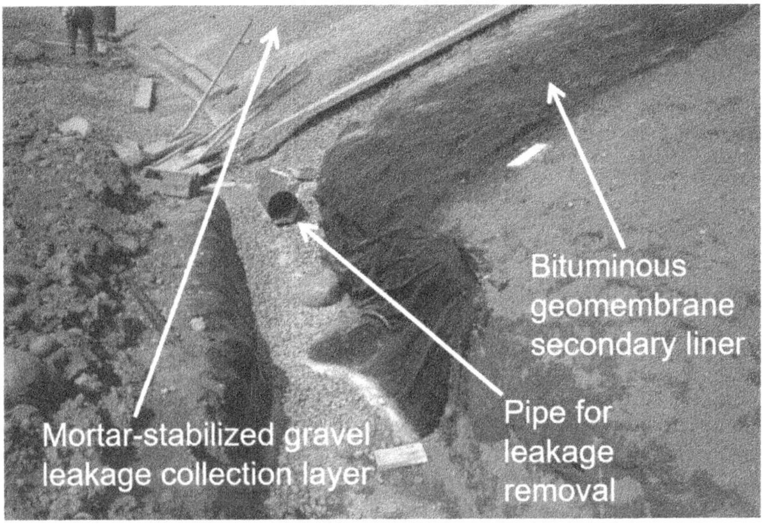

Figure 7.88. *Drainage layer on the secondary liner. For a color version of this figure, see www.iste.co.uk/villard/geosynthetics.zip*

At the foot of the slope are perforated drains with a diameter of 300 mm and a slope of 1%. The drainage is separated into two parts – east and west – to enable any possible leak to be localized. The collector drains have outlets in the monitoring building located at the foot of the reservoir (Figure 7.86).

During the construction, cracks appeared across the slope in the mortar-stabilized gravel layer (Figure 7.89(a)). These were the result of creep movement down the slope. Laboratory tests were carried out at CTGREF (now INREA) to test the ability of the primary sealing layer and its geotextile protection to bridge a crack without tearing.

The primary liner is a 1.5-mm-thick butyl rubber geomembrane installed by the company FIPEC. Butyl is an elastomer, similar to EPDM, highly ductile, elastic and with low ageing from oxidation.

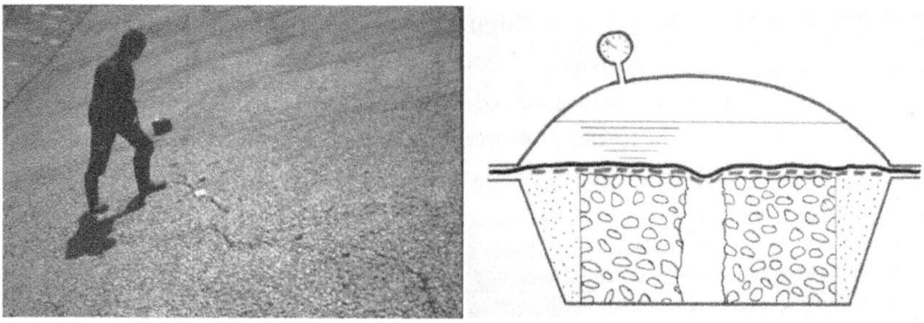

a) b)

Figure 7.89. *(a) Cracks in the stabilized-gravel drainage layer; (b) bridging of the gaps in the drainage layer using the primary liner (laboratory simulation). For a color version of this figure, see www.iste.co.uk/villard/geosynthetics.zip*

The seams are made of heat-vulcanized material (Figure 7.90(a)). The geomembrane is unreinforced except for the 2 m at the head of the slope, where it is reinforced by a woven textile (100 g/m^2) to resist the anchoring forces.

The butyl geomembrane and its underlying geotextile protection are anchored in a trench that is 0.20 m deep and 0.50 m wide, located 0.50 m from the top of the bank (Figures 7.90(b) and (c)).

7.17.3. *Ageing and lessons learnt from the application*

From the time of its construction in 1974, only one incident has been noted: a leak that appeared in 2004. This leak was diagnosed from the flow rate in the collector drain, which demonstrates that the double liner worked well.

It is also noteworthy that the secondary liner had already yielded results just after construction, collecting and removing water for around ten days. However, in this case, the water collected was attributed to condensation of water vapor trapped within the drainage layer, or to water from the mortar itself.

With respect to the leak in 2004, it was localized by means of bubbles that appeared on the surface. This was a leak at moderate depth (around 5 m), very close to the concrete water intake structure, at the seam connecting the geomembrane and the rigid structure (Figure 7.91). The leak was fixed by attaching a geomembrane patch underwater.

a)

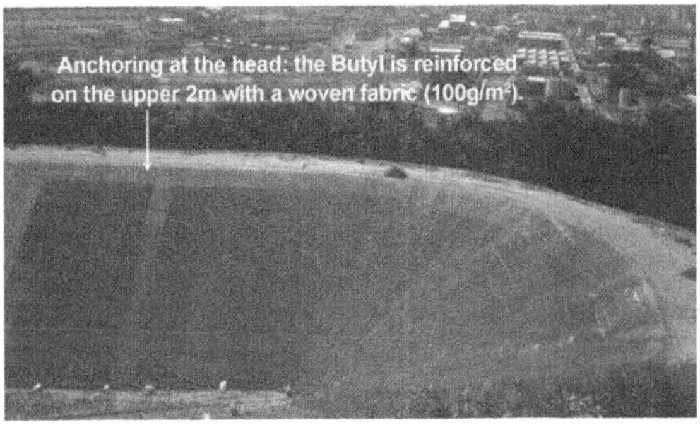

b)

c)

Figure 7.90. *(a) Installation of the primary butyl liner; (b) anchoring of the primary liner: a strip of reinforced butyl rubber geomembrane is used over the top 2 m; (c) after installation of the primary butyl liner (1974). For a color version of this figure, see www.iste.co.uk/villard/geosynthetics.zip*

The structure was visited in 2011 (Giroud and Gourc 2014). The company confirmed that the reservoir had never been completely drained, which obviously minimized the load on the liner. However, the maximum water level is 1 m below the crest and the reinforced butyl in this part was found to have been exposed to UV radiation since 1974 (Figure 7.92).

Figure 7.91. *Locating the leak in 2004 at the level of the water intake (photo from 1974: end of construction). For a color version of this figure, see www.iste.co.uk/villard/geosynthetics.zip*

At the level of the free surface, the butyl geomembrane appeared to be in good condition, without cracks and having retained its extensibility (Figure 7.93).

The surface of the butyl geomembrane was covered by a fine, blackish, oily layer. This is perhaps a carbonaceous compound that migrated to the surface of the geomembrane. The seams also appeared to be in good condition.

Only one hole of around 1 cm^2 was detected on the exposed part to the south of the structure.

To conclude, in this instance, the butyl geomembrane (which is not much in use today, having been replaced by EPDM) has shown good durability. The reinforcing woven on its exposed section (above the water) (Figure 7.90(b)) may have contributed to this behavior.

Figure 7.92. *View of the reservoir in 2011. For a color version of this figure, see www.iste.co.uk/villard/geosynthetics.zip*

Figure 7.93. *Appearance of the butyl geomembrane at the anchoring level in 2011. For a color version of this figure, see www.iste.co.uk/villard/geosynthetics.zip*

7.18. La Hague (1991–1997): cover lining for low-level activity nuclear waste disposal center

7.18.1. *Concept and geotechnology*

The safe storage of nuclear waste is an important global issue. Only the least radioactive waste can be stored in surface storage facilities. Table 7.18 explains where radioactive waste is stored depending on the level of radioactivity and its half-life using the following: VLA (very low level activity), LA (low activity), MA (moderate activity) and HA (high activity).

In France, there are currently three surface storage facilities for nuclear waste. One of these is now closed, since its operational storage phase is over. This is the Manche storage facility (CSM) in La Hague (Figure 7.94), which is constantly monitored. The other two sites are both still active storage facilities. CSM received its first nuclear waste in 1969. When the site shut down in 1994, it had received 527,225 m^3 of waste with a short life and low or moderate radioactivity.

Nuclear waste	Short life (period < 30 years)	Long life (period > 30 years)
Very low activity (VLA) → < 100 Bq/g	Surface storage (Morvilliers VLA-new)	Surface storage (Morvilliers VLA-new)
Low activity (LA)	Surface storage (CSM–closed and Soulaines LMA–active storage site)	Surface storage (research on graphite and radioactive waste)
Moderate activity (MA)	Surface storage (CSM–closed and Soulaines LMA–active storage site)	Deep storage (ongoing on-site research)
High activity (HA)	Deep storage (ongoing on-site research)	Deep storage (ongoing on-site research)

Table 7.18. *Sites for the storage of radioactive waste based on its classification*

Figure 7.94. *CSM at La Hague: general view of the site with vegetated cover-liner (2008). For a color version of this figure, see www.iste.co.uk/villard/geosynthetics.zip*

It is a real challenge to design a cover liner for this kind of a storage site such that it remains effective for 300 years while withstanding mechanical, hydraulic and climatic stresses.

7.18.2. *Design and construction*

The cover liner on the CSM at La Hague is made up of several mineral and synthetic layers (Figure 7.95).

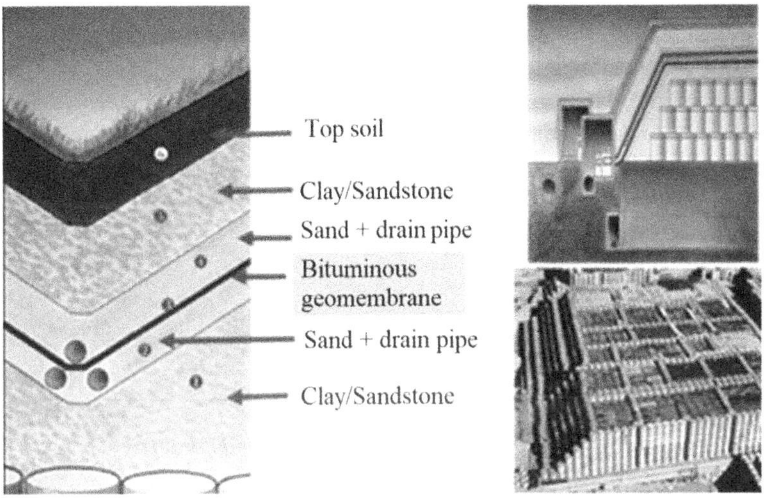

Figure 7.95. *CSM at La Hague: typical cross-section of the cover liner and photo of the storage of the containers. For a color version of this figure, see www.iste.co.uk/villard/geosynthetics.zip*

The surface layer is made up of vegetated soil. It also serves as insulation against climatic stresses. Below this is a layer of clay-sandstone with some very coarse particles. The layer of sand underneath this drains the percolation water, which is collected by a pipe and transported to a basin where quantities are monitored. Under this layer is a bituminous geomembrane, which serves as the sealing layer. Under this geomembrane is a layer of sand that will drain away any possible leakage, which may penetrate the geomembrane. The lower layer of clay-sandstone gives the cover lining its distinctive, "factory roof" shape. This shape was chosen to reduce the path taken by runoff water on the surface in order to minimize infiltration. In a conventional domed storage facility, the water is collected on the sides of the dome.

Low-activity waste is generally stored in blocks or containers of various shapes, and the space between units is filled with sandy material. This kind of storage can lead to possible movement of the units and the creation of gaps between the units, which can then lead to settlement of the cover. The deformation of the cover lining caused by the settlement can lead to cracks developing in the fine soil layer and an increase in the hydraulic conductivity of this layer.

The geomembrane chosen for the CSM site is a bituminous geomembrane GBR-B (Figure 7.96). This kind of geomembrane was chosen by ANDRA (the French national radioactive waste management company) partly because it was the most "natural" geomembrane, even if it involves modified bitumen, and partly because this type of geomembrane can withstand large deformations. Table 7.19 gives a comparison of the extensibility of different types of geomembranes (ε_{max} is the maximum strain in a simple tensile test). The bituminous geomembrane is a geocomposite whose central component is a nonwoven geotextile impregnated with bitumen.

Sealing	Unconfined clay	GBR-HDPE	GBR-C	GBR-B	GBR-PP flexible	GBR-PVC	GBR-EPDM
ε_{max} (%)	0.2–1.5	15	20	> 50	50	300	> 300

Table 7.19. *Maximum deformation of different types of geomembranes in a simple tensile test*

a) b)

Figure 7.96. *CSM at La Hague (1991): (a) construction and storage phase; (b) installation of the bituminous geomembrane. For a color version of this figure, see www.iste.co.uk/villard/geosynthetics.zip*

7.18.3. *Ageing and lessons learnt from the application*

7.18.3.1. *Hydraulic issues*

An attempt has been made to present a hydraulic balance (Gourc et al. 2013), based on the global data from the site since its closure. The quantifiable volumes of water are as follows: precipitation (P), evapotranspiration (ETR), based on data from the National Meteorological Office; there is also runoff (R), collected from the surface of the site cover; the water that infiltrated as far as the (upper) drain, above the upper geomembrane (DuGM); and there is the water that penetrated the geomembrane, collected from the (lower) detection drain (on the lower geomembrane (LGM)) (Figure 7.96(b)).

The calculated (P-ETR-R) value corresponds to the water that may have percolated into the sandy loam (Figure 7.97). This value can be compared to the volume collected in the upper drain plus the volume in the lower drain, (DuGM) + (LGM), which are located above the upper geomembrane and below the lower geomembrane, respectively. If the moisture content of the sandy loam is considered constant over time, then the same value must be found for:

P-ETR-R = DuGM + LGM

Since leaks through the (upper) geomembrane are negligible, (LGM ~ 0) (Figure 7.97), the comparison is restricted to (P-ETR-R) and (DuGM). It can be considered that the difference between the obtained values (Figure 7.97) is acceptable, given that the computation is carried out at the scale of the site, and that the evaluation of the evapotranspiration is only approximate.

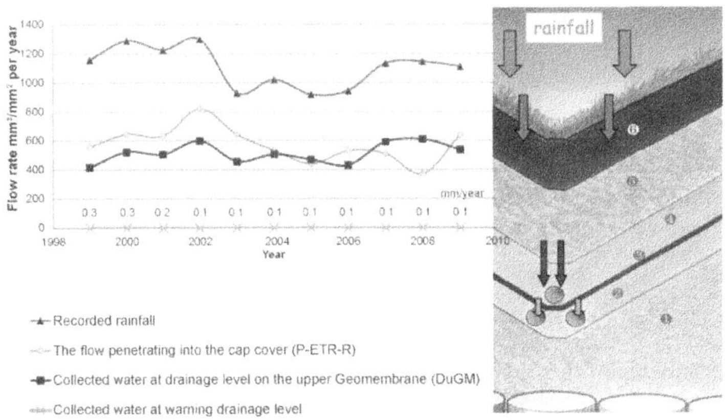

Figure 7.97. *CSM at La Hague: hydraulic balance between 1999 and 2009*

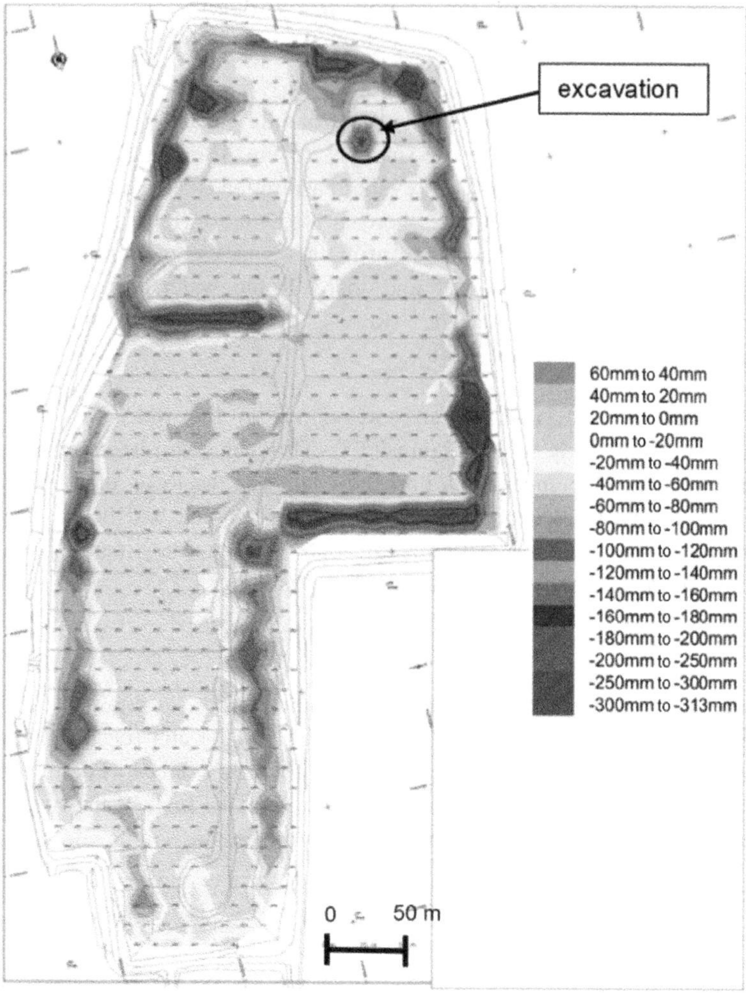

Figure 7.98. *CSM at La Hague, vertical settlement between the end of construction (1998) and 2008 and localization of the zone of differential settlement excavated. For a color version of this figure, see www.iste.co.uk/villard/geosynthetics.zip*

7.18.3.2. *Mechanical issues*

The settlements of the cover lining were recorded as soon as construction of the CSM cover ended in 1998 (Gourc et al. 2013). Figure 7.98 represents the range of settlements between 1998 and 2008. The largest settlements were observed on the slopes lateral to the dome (Ricard et al. 2015).

The settlements on the slope (angle 23°) are related to the slow sliding of the cover over the geomembrane, or to creep of the soil layers (Figures 7.99(a) and (b)). This problem was remedied by reducing the slope angle and by placing an abutment at the foot of the slope.

a) b)

Figure 7.99. *CSM at La Hague: (a) slow sliding over the slope, 20 years after construction. Excavation at the crest of the slope; (b) excavation of the zone of differential settlement (2011). For a color version of this figure, see www.iste.co.uk/villard/geosynthetics.zip*

Furthermore, the top section of the cover had a single zone subject to a progressive, localized collapse (Figure 7.100(a)). In 2011, this zone was excavated in order to analyze this differential settlement (Figure 7.100(b)).

This excavation made it possible to observe the alteration of the covering barrier as a function of its thickness (Figure 7.100). The geomembrane had indeed flexed, but showed no signs of damage (tearing, holes, etc.). The differential settlement appeared to have been caused by the crushing of a waste unit below the geomembrane.

Figure 7.101 represents the iso-elevation curves for the upper geomembrane (the base of the sandy loam layer) and also shows the location of geomembrane samples that were collected, P1 to P4.

If it is assumed that the geomembrane settlements are vertical, the elongation of the geomembrane sheet can be derived: the maximum strain $\varepsilon \approx 11\%$, which is large, but can be tolerated by a bituminous geomembrane.

Burst testing was carried out at INREA on the collected samples to evaluate the loss of tensile strength (Figure 7.101). Diffusion tests, using water with tritium, were carried out at ANDRA in parallel (Marchiol et al. 2015) in order to evaluate the loss of watertightness (Figure 7.102).

Figure 7.100. *CSM at La Hague (2010–2011): (a) site visit close to the zone of differential settlement before excavation (2010); (b) excavation of the zone of differential settlement (2011); (c) excavation of the zone of differential settlement; the geomembrane bridging the collapsed area (2011); (d) collection of samples within the zone of differential settlement (2011). For a color version of this figure, see www.iste.co.uk/villard/geosynthetics.zip*

A complementary structural study was carried out in the 3SR-LTHE laboratories at the Grenoble Alpes University. This studied the influence of a deformation due to stress on the structure and, consequently, on the hydraulic conductivity. It must be recalled that the geomembrane consists of a nonwoven geotextile impregnated with bitumen. Using X-ray tomography made it possible to carry out a preliminary study of the behavior of the bitumen matrix during a tensile strength test, a test that first stresses the nonwoven geotextile before the matrix. Figure 7.103 depicts an example: in a virgin geomembrane sample subject to a deformation of 14% at a temperature of 20°C, a small void can be seen that is doubtless created due to the rotation of fibers with a shearing effect on the bitumen. Since the nonwoven material and the bitumen exhibit complex rheological behavior, new tests need to be developed.

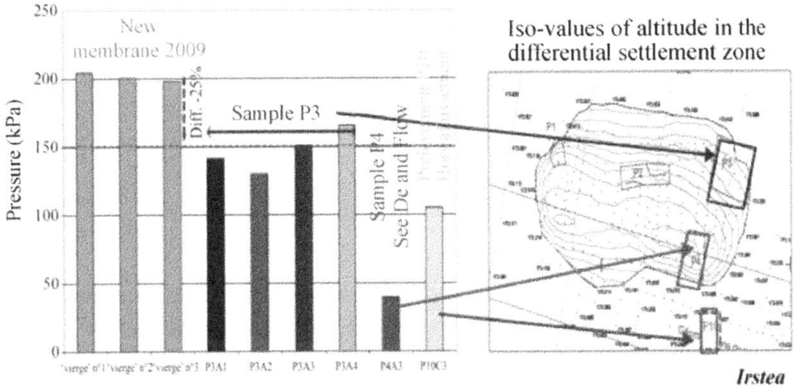

Figure 7.101. *CSM at La Hague: burst testing on the bituminous geomembrane samples collected from different zones. For a color version of this figure, see www.iste.co.uk/villard/geosynthetics.zip*

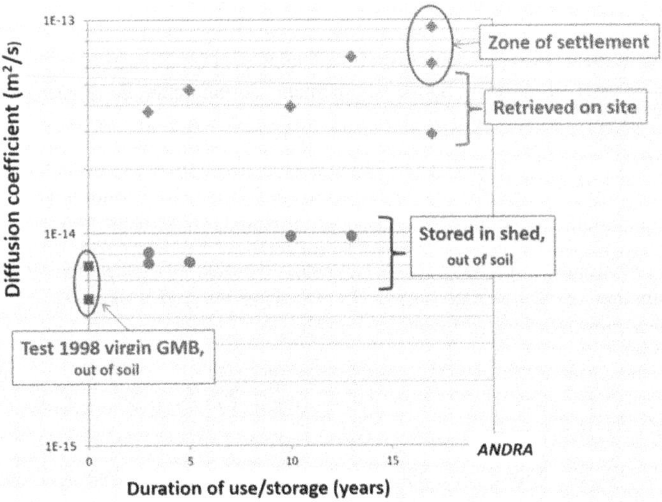

Figure 7.102. *CSM at La Hague: diffusion tests on the bituminous GBR samples collected from different zones. For a color version of this figure, see www.iste.co.uk/villard/geosynthetics.zip*

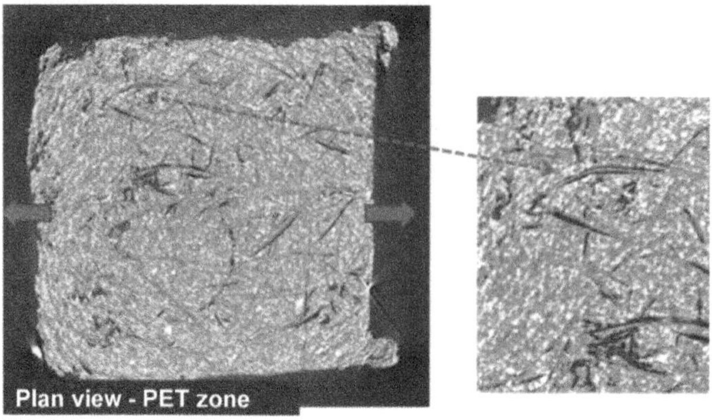

Figure 7.103. *CSM at La Hague: tomography on virgin samples under tensile stress (100 mm/min) (blue arrow: direction of the tensile stress) for an extension* εmax *= 14%; the black shapes are the voids between the PET fibers and the bitumen matrix (3SR-LTHE, 2015)*

7.19. References

AFNOR (2008). XP 84-503, Géomembranes – Dispositifs d'étanchéité par géomembranes (DEG) – Détermination des caractéristiques en traction bi-axiale. Standard, AFNOR, Paris.

AFNOR (2017). NF G38-061. Utilisation des géotextiles et produits apparentés – Systèmes de drainage et de filtration – Dimensionnement et éléments de conception. Standard, AFNOR, Paris.

AFNOR (2020). XP G38-065. Géosynthétiques, géotextiles et produits apparentés – Renforcement de la base des remblais sur zones à risques d'effondrement – Justification du dimensionnement et éléments de conception. Standard, AFNOR, Paris.

Alonso, E., Degoutte, G., Girard, H. (1990). Results of seventeen years of using geomembranes in dams and basins. In *4th International Conference on Geotextiles Geomembranes and Related Products*, The Hague.

Blivet, J.C. and Gourc, J.P. (1998). Large retaining structures reinforced by geosynthetics: French case histories. In *Fourth International Conférence on Case Histories on Geotechnical Engineering*, St. Louis, MO.

Borie, F., Velly, B., Montgenet, Ph., Perrier, H., Simon, A. (1991). Lutte contre les effondrements karstiques. *Revue générale des routes*, 690, 61–63.

Carbone, L., Gourc, J.P., Caroubba, P., Pavanello, P., Moraci, N. (2015). Dry friction behaviour of a geosynthetic interface using inclined plane and shaking table tests. *Geotextiles and Geomembranes*, 43(4), 293–306.

Cartier, G. (1985). Stabilisation des pentes instables par clouage. In *Journées des Laboratoires des Ponts et Chaussées*, Lyon.

Cassard, A., Kern, F., Mathieu, G. (1979). Utilisation des techniques de renforcement dans les barrages en terre. In *Colloque inter. sur le renforcement des sols*, Paris.

Cazzuffi, D., Giroud, J.P., Scuero, A.,Vaschetti, G. (2010). Geosynthetic barriers systems for dams. In *9th International Conference on Geosynthetics*. Keynote Lecture, Guarujá.

Cazzuffi, D., Moraci, N., Mandaglio, M.C., Ielo, D. (2016). Evolution in design of geotextile filters. In *EuroGeo 6*, Ljubljana.

CFG (2016). Drainage et filtration, recommandations pour l'emploi des géosynthétiques dans les systèmes de drainage et de filtration [Online]. Available at: https://www.cfg.asso.fr/publications/guides-de-recommandations/recommandations-pour-lemploi-des-geosyntheti ques-dans-les.

Delmas, P. (1988). Ouvrages de soutènement renforcés par géotextiles : l'expérience du réseau des LPC. *Bulletin de liaison des ponts et chaussées*, 153, 89–101.

Delmas, P. and Matichard, Y. (1986). Landslide confortation with geotextile reinforced earth work. In *ICG 3 International Conference on Geotextiles*, Vienna.

Delmas, P., Puig, J., Schaeffner, M. (1984). Procédé et dispositif pour la construction de massifs de sols renforcés, notamment par nappes de géotextiles. Patent no. 2569741.

Delmas, P., Berche, J., Gourc, J.P. (1986). Le dimensionnement des ouvrages renforcés par géotextiles : programme Cartage. *Bulletin de liaison des ponts et chaussées*, 142, 33–44.

Delmas, P., Faure,Y. H., Farkouh, B., Nancey, A. (1994). Long term behaviour of a geotextile as a filter in a 24 years old earth dam: Valcros. In *Proceedings of the Fifth International Conference on Geotextiles, Geomembranes and Related products*. International Geotextile Society, Singapore.

Delmas, P., Villard, P., Huckert, A. (2015). Short and long term design of geosynthetic reinforced structure over soil subsidences: Taking into account safety. In *10ème Rencontres Géosynthétiques*, La Rochelle.

Faure, Y.H., Farkouh, B., Meriaux, P. (1994). Long term behaviour of trench drain in the Alps. In *5th International Conference on Geotextiles Geomembranes and Related Products*, Singapore.

Faure, Y.H., Kehila, Y., Farkouh, B. (1996). Behaviour of experimental in-situ trench drains in varved clays. In *Geofilter 96*, Montreal.

Faure, Y.H., Farkouh, B., Delmas, P., Nancey, A. (1999). Analysis of geotextile filter behaviour after 21 years in Valcros dam. *Geotextiles and Geomembranes*, 17(5–6), 353–370.

Girard, H., Fischer, S., Alonso, E. (1990). Problems of friction posed by the use of geomembranes on dam slopes – Examples and measurements. *Geotextiles and Geomembranes*, 9, 129–143.

Giroud, J.P. (1973). L'étanchéité des retenues d'eau par feuilles déroulées. *Annales ITBTP*, 312(TP 161), 94–112.

Giroud, J.P. (1985). Filter criteria for geotextile. *Geotextiles and Geomembranes*, 2(3), 103–108.

Giroud, J.P. and Gourc, J.P. (2014). The first double geomembrane liner forty years later. In *10th International Conference on Geosynthetics*, Berlin.

Giroud, J.P., Gourc, J.P., Bally, P., Delmas, P. (1977). Comportement d'un textile non-tissé dans un barrage en terre. In *C.R. Call. Int. Sols Textiles*, Paris.

Gourc, J.P. (1982). Quelques aspects du comportement des géotextiles en mécanique des sols. PhD Thesis, Université de Grenoble, Grenoble.

Gourc, J.P. (1996). Retaining structures with geosynthetics: A mature technique but with some questions pending. In *1st European Geosynthetics Conference on Eurogeo 1*. Keynote Lecture, Maastricht.

Gourc, J.P. and Matichard, Y. (1994). Two decades of geosynthetic reinforced retaining structures in France. In *Recent Case Histories of Permanent Geosynthetic Reinforced Soil Retaining Walls*, Tatsuoka, F. and Leshchinsky, D. (eds). Balkema, Rotterdam.

Gourc, J.P., Ratel, A., Delmas, P. (1986). Design of fabric reinforced retaining walls: The displacement method. In *3rd International Conference on Geotextiles*, Vienna.

Gourc, J.P., Arab, R., Giraud, H. (2001). Calibration and validation of design methods for geosynthetic-reinforced retaining structures using partial factors. *Geosynthetics International*, 8(2), 163–191.

Gourc, J.P., Verstaevel, M., Rey, D. (2013). Spécificités des couvertures de stockage de déchets faiblement radioactifs. In *9th Rencontres 13*, Dijon.

Hérault, A., Mieussens, J.C., Nancey, A., Dalbion, J., Delmas, P. (1995). The geotextiles in reinforcement of retaining wall: A well tested technique – wall WS13 RN20 Foix-Tarascon sur Ariège. In *Rencontres geosynthetiques 95*, Beaune.

ICOLD (2010). Geomembrane sealing system for dams – Design principles and return of experience. In *Bulletin 135, International Commission of Large Dams-Comité International des Grands Barrages*, Paris.

Kern, F. (1977). Réalisation d'un barrage en terre avec parement aval vertical au moyen de poches en textiles. In *Colloque Inter. "Sols et Textiles"*, Paris.

Khay, M. and Claquin, Y. (2000). Reinforced soil-thread technique in protective mounds of pressure tank. Document, Geoasia, Kuala Lumpur.

Khay, M., Gigan, J., Ledelliou, M. (1990). Reinforcement with continuous threads: Technical developments and design method. In *4th International Conference on Geotextiles Geomembranes and Related Products*, The Hague.

Koffler, A. (1995). Renewed impermeabilization of canal de Jonage. In *Rencontres géosynthétiques 95*, Beaune.

Lafleur, J. (1999). Selection of geotextiles to filter broadly graded cohesionless soils. *Geotextiles and Geomembranes*, 17, 99–312.

Lambert, S. and Touze-Foltz, N. (2000). A test for measuring permeability of geomembranes. In *2nd European Conference on Geosynthetics*, Bologna.

Lavaud, S., Vouyovitch Van Schoors, L., Barberis, N., Lozach, D., Khay, M. (2011). Evolution des caracteristiques des fibres polyester d'un remblai en Texsol après 20 ans. In *8th Rencontres géosynthétiques*, Tours.

Leflaive, E. (1982). The reinforcement of granular material with continuous fibers. In *Second International Conference on Geotextiles*, Las Vegas.

Leflaive, E. (1988). Texsol, already more than fifty successful applications. In *International Geotechnical Symposium on Theory and Practice of Earth Reinforcement*, Fukuoka.

Marchiol, A., Ricard, D., Robinet, J. (2015). Surveillance d'une géomembrane bitumineuse oxydée dand une couverture de stockage sur le long terme. In *10th Rencontres géosynthetique*, La Rochelle.

Mitchell, J. and Villet, W. (1987). Reinforcement of earth slopes and embankments. Report, National Cooperative Heighway Research Progarmme, Transportation Research Board, Washington, D.C.

Perrier, H., Ducol, J., Simon, A., Lacroix, G. (1990). Renforcement des routes par géotextile: prévention contre les effondrements karstique. In *4th International Conference on Geotextiles, Geomembranes and Related Products*, The Hague.

Puig, J. and Blivet, J. (1973). Remblai à talus vertical armé avec un textile synthétique. *Bulletin des laboratoires des ponts et chaussées*, 64, 13–18.

Risseeuw, P. and Alexiew, D. (1999). *Erfahrungen mit einer der ältesten Geotextilbewehrten Stützwände in Europa: Bauweise. Statik, Verhalten, Langzeitständigkeit.* Deutsche Gesellschaft fur Geotechnik e.V, Essen.

Risseeuw, P. and Schmidt, H. (1990). Hydrolysis of HT polyester yarns in water at moderate temperatures. In *4th International Conference on Geotextiles, Geomembranes and Related Products*, The Hague.

Royet, P., Girard, H., Poulain, D., Tisserand, C. (2002). Comportement des géomembranes dans les barrages français en remblai. *Ingéniéries-E.A.T.*, 31, 3–15.

Touze-Foltz, N., Croissant, D., Farcas, F., Royet, P. (2010). Quantification of oxidized bituminous geomembrane ageing through hydraulic testing. In *9th International Conference on Geosynthetics*, Guarujá.

Veldhuizen Van Zanten, R. (1986). *Geotextiles and Geomembranes in Civil Engineering.* Balkema, Rotterdam.

Veylon, G., Stoltz, G., Mériaux, P., Faure, Y.H., Touze-Foltz, N. (2016). Performance of geotextile filters after 18 years' service in drainage trenches. *Geotextiles and Geomembranes*, 44 , 515–533.

List of Authors

Philippe DELMAS
Formerly Cnam
Paris
France

Fabienne FARCAS
MAST-CPDM
Université Gustave Eiffel
Marne-la-Vallée
France

J.P. GIROUD
Consulting Engineer and Expert
Paris
France

Jean-Pierre GOURC
Formerly Polytech
Grenoble IGE
Université Grenoble Alpes
France

Audrey HUCKERT
Laboratoire 3SR
Université Grenoble Alpes
France

Guillaume STOLTZ
RECOVER
INRAE
Aix-Marseille Université
Aix-en-Provence
France

Francis TANO
MERAMO
Chatenay-Malabry
France

Nathalie TOUZE
SDAR
INRAE
Université Paris-Saclay
Jouy-en-Josas
France

Laetitia VAN SCHOORS
MAST-CPDM
Université Gustave Eiffel
Marne-la-Vallée
France

Pascal VILLARD
Laboratoire 3SR
Université Grenoble Alpes
France

Index

E, F

G, H

I, L

M, N

O, P

Q, R

S, T

V, W

Printed and bound by CPI Group (UK) Ltd, Croydon, CR0 4YY

10/07/2025

14701701-0001